With everlasting love, to my wife, Marji, and son, Trevor,
for giving new and special meaning to my life.

William C. Whiting

To my late mother, Barbara, for giving me the gift of life and teaching me
how to love it, live it, and share it. To my late father, George, for always
believing in me and teaching me how to turn challenges into opportunities.
And to my stepmother, Nadine, for her continual love, guidance, and support.

Stuart Rugg

contents

preface

No earthly creation is as intricate, versatile, or mysterious as the human body. Our bodies perform countless functions that ensure our existence; and of all the body's functions, movement is arguably the most essential, for without movement, the human body could not survive. With movement, the body flourishes. At one end of the movement spectrum, the capacity for purposeful movement allows us to maintain fundamental physiological processes; at the other end, it allows us to pursue the limits of athletic and artistic expression.

For those involved in the many areas of human health and performance, knowledge of the body's structure and function is essential. Students in these areas undoubtedly have taken introductory courses in human anatomy. All too often, however, these courses (which attempt to cover all of the body's systems in a single academic term) provide students with the basics of *structural anatomy* but offer limited exposure to the elegance and complexity of the body's *functional movement anatomy.*

> If anything is sacred, the human body is sacred.
>
> — *Walt Whitman*
> *(1819-1892)*

From our experience teaching human anatomy, kinesiology, and biomechanics over the past two decades, we can attest to the unfortunate fact that many students emerge from their introductory anatomy experience with limited competency in applying their anatomical knowledge to human movement problems. These students have learned what we not-too-euphemistically, and only somewhat kiddingly, refer to as "dead anatomy."

The purpose of this book is to bring anatomy to life by exploring the marvelous potential of the human body to express itself through movement and to provide students with the information and skills they need to appreciate and assess *dynamic human anatomy,* or **dynatomy.**

This book has been developed primarily for students who already have taken, or currently are taking, an introductory course in human anatomy and who need a more detailed exposure to concepts of human *movement* anatomy. These students include those in the fields of exercise science and human movement studies, kinesiology, biomechanics, physical education, coaching, athletic training, ergonomics, and the health sciences (e.g., medicine, physical therapy).

The text is accompanied by a CD, *Essentials of Interactive Functional Anatomy* (Primal Pictures), which depicts human musculoskeletal anatomy and movement in intricate detail. As a companion to *Dynatomy,* the CD provides students and instructors with a state-of-the-art three-dimensional model of the human body. *Essentials of Interactive Functional Anatomy* also includes interactive functions and advanced animations that bring to life the movement concepts presented in the text. We have indicated the places in the text for which it is especially worthwhile to have a look at the CD by placing an icon in the margin.

Given the amount of material presented and memorized in introductory anatomy courses, it is neither unusual nor unexpected that students forget many of the details. Thus part I (Anatomical Foundations and Essentials of Movement Control) provides a concise review of relevant anatomical information and neuromechanical concepts. Part I begins (chapter 1) with an introduction to the dynamics of human movement and essentials of anatomical structure. The chapter presents a brief overview of motor behavior, including life-span motor development, motor control, and motor learning, and establishes a contextual framework for later discussion of specific movements and populations.

Chapter 2 (Osteology and the Skeletal System) describes the microstructure and macroscopic morphology of bone and the organization of the skeletal system as well as identifies specific bones and bony landmarks.

Part I continues in chapter 3 (Joint Motion and the Articular System) with a discussion of joint structure, function, and motion, along with a description of major joints of the extremities and spine. Chapter 4 (Myology and the Muscular System) begins with discussion of the structure and function of skeletal muscle and the basic physiology and mechanics of muscle action. The chapter continues with exploration of muscle actions; voluntary, reflex, and stereotypical movements; nervous system control of muscle action; and factors (e.g., muscle length and velocity, fiber type, and muscle architecture) that affect muscle force production. Part I concludes in chapter 5 (Muscles of Movement) with a presentation of the primary muscles responsible for the production and control of human movement.

With the requisite anatomical and neuromechanical foundations in place, part II (Applied Dynatomy) provides the essentials of the dynatomical approach to movement. This part begins in chapter 6 (Mechanics of Movement) with a review of mechanical concepts essential to an understanding of human movement. Chapter 7 (Muscular Control of Movement and Movement Assessment) explains the *muscle control formula*, a set of simple steps used to identify the muscles and types of muscle actions responsible for producing or controlling any human movement. Numerous examples, using simple movement patterns, demonstrate the utility of the formula. Mastery of the muscle control formula provides a valuable tool for independent assessment of muscle action across a broad spectrum of human movements. Chapter 7 concludes with consideration of several important topics relevant to movement assessment. These include single-joint versus multijoint movements, coordination, movement efficiency, and assessment of movement.

Chapter 8 (Fundamentals of Posture, Balance, and Walking) examines the dynatomy of posture and balance as well as the most common movement form, walking. Chapter 9 (Fundamentals of Running, Jumping, Throwing, Kicking, and Lifting) continues with a discussion of everyday movement tasks. Chapter 10 (Analysis of Exercise and Sport Movements) extends the application of the dynatomical approach to movement patterns in resistance training, cycling, swimming, and dance. Chapter 11 (Future Directions of Human Movement Studies) concludes with a brief discussion of advances in medicine and technology that affect the study of human movement, demographic and social trends that may create movement challenges in the future, and a provocative note on the limits to human performance.

One of our primary goals has been to include concepts not found in many traditional anatomy texts, concepts that emphasize function and application. We are confident that an understanding of the material covered will enhance your preparation for work in the many areas of human movement. To assist you, we included a comprehensive glossary that defines hundreds of important terms and concepts. Terms bolded in the text can be found in the glossary. Each chapter also includes objectives and critical thinking questions to guide and challenge students, as well as suggested readings for further study.

In addition to the considerable amount of information presented, the text is sprinkled with the words of others because we agree with British politician Benjamin D'Israeli, who wrote that "the wisdom of the wise, and the experience of ages, may be preserved by quotation" (D'Israeli, 2005). Throughout the book, we include quotations by many famous, and some not-so-famous, persons. As Ralph Waldo Emerson noted, "By necessity, by proclivity, and by delight, we all quote" (Emerson, 2005).

One final perspective note: Please keep in mind that the material in this text is only a beginning. Consideration of any one of the many topics presented could be expanded to fill a chapter, or even a book, of its own. We make no pretense of having comprehensively covered any of the topics but rather present a concise introduction to the fascinating area of human movement.

This book will have served its purpose well if the reader emerges with an appreciation for human movement that is more wondrous and less mysterious than when the journey began. In the words of British author Laurence Sterne, "So much of motion, is so much of life, and so much of joy." (Sterne 1980, p. 354)

acknowledgments

While only two names appear on its cover, this book could not have been completed without the contributions of many people. We extend our thanks to the many friends and colleagues who have contributed to its completion. We give particular thanks to Dr. Robert Gregor and Dr. Ronald Zernicke for sharing their knowledge and enthusiasm for biomechanics and helping us develop our love of teaching. Special thanks to Dr. Judith Smith for her pioneering work in establishing the human anatomy program, in the then-Department of Kinesiology at UCLA, that helped provide us with our foundation in anatomy and our belief in the remarkable power of human movement.

Individually, Dr. Whiting gives special thanks to two close friends and professional colleagues, Dr. George Salem and Dr. Steven Loy, for always being there to lend a helping hand or an understanding ear. Dr. Rugg gives special thanks to his colleague, Dr. Lynn Mehl, for believing in him and being such a great team player, and to his colleague and co-researcher, Dr. Eric Sternlicht, for his continual support, guidance, and great friendship.

We thank the staff at Human Kinetics, most especially Dr. Loarn Robertson and Renee Thomas Pyrtel, for their patience, support, and belief in this project.

And finally, and most important, we thank our families for their unwavering support and love during the course of this project, and always.

part I

Anatomical Foundations and Essentials of Movement Control

The five chapters of part I present a concise review of human anatomy and a conceptual framework for our discussion of human movement. Chapter 1 introduces anatomical and movement concepts that form the foundation for understanding details presented in subsequent chapters. Chapter 2 reviews the organization of the skeletal system and the details of bone structure and function. Using the information presented in chapter 2, we continue in chapter 3 with discussion of the articular system and joint motion, including review of the major joints involved in human movement.

Chapter 4 explores the structure and function of skeletal muscle, the tissue that provides the force needed to produce and control our movements. In this chapter we also examine factors that affect muscle force production. Chapter 5 concludes part I with a presentation of the muscles of movement and a summary of the specific muscles responsible for movement control.

The information presented in part I is essential to our understanding of dynamic human anatomy and sets the stage for our application of the dynatomy approach to a wide variety of human movements in part II (Applied Dynatomy).

one

Introduction to Human Anatomy and Movement

objectives

After studying this chapter, you will be able to do the following:

- Define the terms *anatomy* and *physiology* and describe the relationship between these two areas of study

- Explain the importance of movement in our daily lives

- Understand how movement changes across the life span

- Describe the importance of understanding motor behavior and its subareas of motor control, motor learning, and motor development

- Explain movement considerations in younger and older populations

- Describe the importance of differences in movement ability among individuals

- Explain the anatomical concepts of complexity, variability, individuality, adaptability, connectivity, and asymmetry

- Describe the levels of structural organization and primary tissue types

- Explain the structure and function of muscle and connective tissues

- Define and explain anatomical terms describing body regions, body positions, movement planes, axes of rotation, joint positions, and movements

- Appreciate the multidisciplinary perspective necessary to understand human movement

> Considering all of the marvelous tools of one kind or another that humans have invented during recent decades (and before, for that matter) one is struck by the fact that none of them is quite so complex, potentially perfect, or wondrous as the human body, the one possession we have in common.
>
> —Isaac Asimov (1920-1992)

n this introductory chapter, we establish the conceptual framework for understanding the details of human anatomy and movement in the chapters that follow. In building this framework, we take a life-span approach and focus on general anatomical terminology and concepts and an overview of human structure. We emphasize the elements of anatomy (e.g., bones and muscles) directly related to human movement.

Introduction to Human Anatomy

Anatomy, simply defined, is the study of the structure of organisms. The term is derived from the Greek *anatome,* which means "to dissect" or "to cut." The organism under consideration in this book is the human. Although dissection historically has been a primary means of discovery about the structure of the human organism, and it remains a useful educational tool in training professionals in the health sciences (e.g., medicine, physical therapy), in recent decades, technological advances of many kinds have enhanced our understanding of the body's structure without our needing to rely exclusively on cutting it open.

Anatomy is an expansive science that encompasses many branches, or subdivisions. Some of these subdivisions are described in table 1.1. In establishing our anatomical foundation for study of human movement in part I of this book, we focus on gross (macroscopic) anatomy, systemic anatomy (especially the skeletal, muscle, and nervous systems), regional anatomy as it relates to specific movement patterns, and most important, functional anatomy.

Although anatomy is the focus of our study, the related area of **physiology,** which deals with the *functions* of the body parts and systems, must also be mentioned. Anatomy and physiology are closely related, and study of one without appropriate consideration of the other makes little sense. We begin by presenting, in elementary terms, the origin, development, and structure of the body's components and then show how the parts are both structurally and functionally interconnected.

Table 1.1 Subdivisions of Anatomy

Anatomy subdivision	Description
Gross (macroscopic) anatomy	Study of structures without the aid of a microscope
Microscopic anatomy	Study of structures using microscopic techniques
Systemic anatomy	Study of specific body systems (e.g., skeletal, nervous, respiratory, cardiovascular)
Regional anatomy	Study of specific body regions (e.g., head, extremities)
Functional anatomy	Study of how body systems (e.g., skeletal, muscle, nervous) cooperate to perform various functions
Surface anatomy	Study of landmarks on the surface of the body
Developmental anatomy	Study of structural development across the life span, from fertilization to death
Embryology	Study of development from fertilization through the 8th week in utero
Histology	Microscopic study of tissue structure
Cytology	Microscopic study of cell structure

Vesalius

Some of the most admired anatomical treasures of the Renaissance were drawn by Vesalius (1514-1564). His master work, *De Humani Corporis Fabrica* (1543), stands as one of the great volumes in the history of modern science. In this work, Vesalius blended text with picture into a truly integrated presentation of his understanding of human anatomy (see figure 1.1).

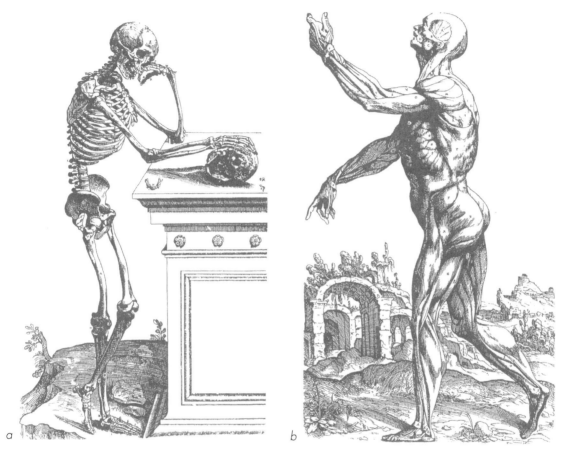

a   b

Figure 1.1   The anatomical legacy of Vesalius.

This approach to the study of functional anatomy is by no means a novel one. Leonardo da Vinci (1452-1519), perhaps the best-known figure of the Renaissance, wrote the following regarding his plan for a "Treatise on Anatomy," a work that never reached completion:

> This work should begin with the conception of man and describe the nature of the womb, and how the child lives in it, and up to what stage it dwells there, and the manner of its quickening and feeding, and its growth, and what interval exists between one stage of growth and another, and what drives it forth from the body of the mother, and for what reason it sometimes comes forth from the belly of its mother before the proper time.

> Then describe which are the members which grew more than the others after the child is born, and give the measurements of a child of one year.

Next describe a grown male and female, and their measurements, and the nature of their complexions, color and physiognomy. Afterwards describe how he is composed of vessels, nerves, muscles and bones. This you will do at the end of the book. (O'Malley & Saunders, 1997, p. 31)

We begin with an overview of human movement, followed by an introduction to concepts of functional anatomy. These anatomical concepts form a necessary foundation for your understanding of dynamic human anatomy.

Introduction to Human Movement

Movement is a fundamental behavior essential for life itself. Life processes such as blood circulation, respiration, and muscle contraction require motion, as do activities such as walking, bending, and lifting. The human organism seeks, consciously or not, to move. Children, in particular, provide clear evidence of the inherent nature of humans to move. They never seem to stop. Even as we age and slow down, movement remains an essential part of our lives. As noted by French scientist and philosopher Blaise Pascal (1623-1662), "Our nature consists in motion; complete rest is death" (Pascal & Krailsheimer, 1995).

> From the invisible atom to the celestial body list in space, everything is movement. . . . It is the most apparent characteristic of life; it manifests itself in all functions, it is even the essence of several of them.
>
> —Etienne-Jules Marey (1830-1904)

At one end of the spectrum, the capacity for purposeful movement supports basic physiological processes; at the other end, movement allows us to explore the limits of athletic and artistic expression. Limited movement, such as when a person is bedridden or elects a sedentary lifestyle, can contribute to adverse health conditions such as cardiovascular disease, diabetes, and cancer. Thus our ability to move, or the choice to limit movement, may contribute, either directly or indirectly, to our susceptibility to disease and injury.

The study of human movement across all of its dimensions is known as **kinesiology** (Gk. *kinesis,* fr. *kinein* "to move"). Kinesiology is a broad discipline that encompasses both the science and art of human movement. It draws from many related disciplines, including anatomy, physiology, biomechanics, motor behavior, and psychology, together with clinical and applied disciplines such as medicine, physical therapy, engineering, and physical education. A primary goal of studying kinesiology is to identify the underlying mechanisms and consequences of human movement.

The range of human movements is broad and complex. Some movements, such as throwing, consist of a single episode, while others, such as walking, involve repetitive movement cycles. And none of us moves in the same way, even while doing the same thing. Each person, for example, adopts a unique walking pattern based on individual structure, purpose, and style. This can be demonstrated in the common experience of seeing a person's silhouette in the distance and knowing who it is "by the way he walks" long before we recognize him by facial or body features. Something in the way he moves tells us who he is.

Movement Across the Life Span

> Myself now and myself a while ago are indeed two; but when better, I simply cannot say. It would be fine to be old if we traveled only toward improvement.
>
> —Michel de Montaigne (1533-1592)

As we navigate life's journey, each of us becomes aware of our own movement capabilities and limitations. As children, we progress from immobility to crawling, walking, running, and jumping. With age, we grow larger and stronger, and we typically become more proficient in our movement patterns throughout most of our lives. Our later years may be characterized by movement pattern changes due to muscle strength decreases, fatigue, nervous system declines, postural changes, injury, disease, and environmental factors. One thing is clear—our ability to move changes across the life span. These changes can enhance movement potential, as when a child grows and develops, or limit movement, as in the case where injury, disease, or age-related functional declines cause a deterioration in movement capacity.

The Legacy of Borelli

In the history of science, the name Giovanni Alfonso Borelli (1608-1679) does not usually appear on lists with such giants as Galileo and Sir Isaac Newton. In the history of movement science, however, Borelli deserves a prominent place. His definitive treatise, *De Motu Animalium*, serves as the seminal work on the science of animal movement. "This book . . . postulates that every function in the living body, animal or vegetable, manifests itself through movement: macroscopic and apparent, as in locomotion, or microscopic, on an atomic dimension, as the movement in which atoms come in contact to form living matter. Borelli's premise and fundamental aim was to integrate physiology and physical science" (Cappozzo & Marchetti, 1992, pp. 33-35).

This richly illustrated volume (see figure 1.2) includes discussion of musculoskeletal anatomy and muscle physiology, along with application of mechanical principles to movement across a broad spectrum of animal motion. Borelli modeled the leverage of muscles, tendons, and bone and applied such mechanical constructs to the analysis of human gait, as well as to the locomotor patterns of horses, insects, birds, and fish.

In many ways, Borelli was ahead of his time. He was restricted—by limited technology and yet-to-be-formalized theories of mechanics—from fully pursuing the concepts he clearly embraced. Regrettably, Borelli never saw his *De Motu Animalium* in published form. Most of Borelli's work was published only after his death (Cappozzo & Marchetti, 1992).

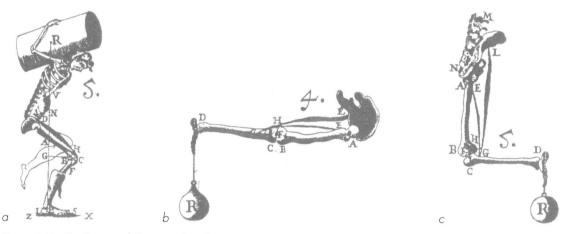

Figure 1.2   The legacy of Giovanni Borelli.

Anatomical changes that occur from day to day are barely detectable, if at all, but as days turn to weeks, weeks to months, and months to years, the changes become evident. These changes may be due to growth (e.g., an adult is larger and stronger than a child); physical conditioning (e.g., an athlete performs better after a strength or endurance training program); injury or disease (e.g., the growth of a child's bone may be affected by a serious fracture); or a variety of environmental, sociological, psychological, or cultural influences.

Each of us changes in countless unique ways throughout our lives. These changes directly affect our potential and capacity to move. The chapters to follow take a life-span approach to functional anatomy and human movement and show many ways in which movement is affected by our time and place along life's journey. The acquisition and refinement of movement skills across the life span are examined through the study of motor behavior and its subareas: motor control, motor learning, and motor development.

Motor Behavior

In the context of human movement, the term **motor** describes things related to or involving muscular movement. **Motor behavior** is an umbrella term that covers several specialized areas related to muscular control of movement: motor control, motor learning, and motor development. Each contributes to our understanding of human movement.

Each area of motor behavior typically encompasses a distinctive time frame, with motor control covering events with very short time intervals; motor learning involving times of hours, days, and weeks; and motor development addressing events over months, years, and even decades. For example, to control the muscles involved in throwing a ball, the nervous system sends signals to the shoulder and arm. These signals may last for a fraction of a second. To learn the task of throwing may take days or weeks to accomplish. And to see developmental changes in throwing from childhood to adulthood, we need to observe the task across many years. Though these concepts are clearly related, we consider each in turn.

Motor Control

Motor control is the study of the neural, physical, and behavioral aspects of movement (Schmidt & Lee, 1998). More specifically, motor control refers to how the body's systems organize and control muscles involved in movement. The system primarily involved in control is the nervous system, which consists of the brain, spinal cord, and peripheral nerves emerging from the cord.

The degree of motor control clearly depends on the developmental, or maturational, stage of the individual. A 2-year-old child, for example, cannot be expected to have the same degree of movement control as a 10-year-old or an adult. As the systems involved in movement (nervous, muscular, skeletal, cardiovascular, respiratory) develop, the body is better able to control movements and develops a wider variety of potential movement skills.

The study of motor control originated with neurophysiologists who explored how the nervous system worked with muscles to produce and control movement. Over the decades, various models have attempted to explain the mechanisms of motor control.

Motor Learning

Experience and the environment can teach us a lot. So too can others who share their knowledge and experience to help us learn. **Motor learning** deals with how we learn to move or to develop motor skills. A **motor skill** is defined as a voluntary movement used to complete a desired task action or achieve a specific goal. When we first begin to learn a new motor skill, our movements usually are awkward, uncoordinated, and inefficient. With practice, we learn to refine the movement until it becomes relatively permanent and proficient.

The process of motor learning involves several stages, or phases. These phases, as first presented by Fitts (1964), are the cognitive phase, associative phase, and autonomous phase. In the initial cognitive phase, a person must devote considerable conscious thought to the movement task and tries various strategies. With repetition, the mover retains the strategies that best accomplish the task and discards the movements that do not. In the associative phase, movements are less variable, and the mover determines the best movement strategy. The movement requires less cognitive involvement, and the mover can focus on refining and perfecting the movement. In the final autonomous phase, the movement becomes automatic, or instinctive, and the mover can attend to other factors such as environmental conditions that may affect task performance and movement.

Motor Development

Motor development explores the changes in movement behavior that occur as one progresses through the life span from infancy until death. The changes typically are continuous, sequential, and age related. An infant, for example, increases her mobility by progressing from crawling and creeping to walking, and then from walking to running. These developments occur continuously, in a predefined sequence, and are obviously age related.

Movement changes are directed by numerous individual, environmental, and task constraints (Newell, 1986). Individual constraints may be either structural (e.g., body size and strength) or functional (e.g., motivation). Environmental constraints include elements outside the body and can be physical (e.g., weather or surface conditions), sociological, or cultural (e.g., societal norms or traditions that favor one group's participation over another's). Task constraints refer to external factors inherent to the task at hand. A participant in a wheelchair basketball game has obvious constraints imposed by his own wheelchair and those of others in the game, along with the challenges of avoiding his opponent's defensive efforts.

Movement Considerations in the Young

Comparison of movement between younger and older persons readily shows that children are *not* just miniature adults. Infants and children are anatomically, physiologically, psychologically, and emotionally different from adults, and their structure and movement behaviors reflect these differences. For example, infants' body proportions differ from adults', as can be seen in the fact that infants have proportionally shorter limbs and larger heads than adults do. This difference in weight distribution plays an important role in movement dynamics of infants compared with older children and adults.

There are near-universal similarities in human developmental progression. However, the progressive changes do not occur at the same age, or at the same rate, for every individual. Nor are the developmental changes the same for everyone. Averages exist across a given population, but within the population considerable variability is apparent. As expected, average values are reported for developmental measures, but many people fall above or below the average. The individual nature of development should always be kept in mind.

> The child amidst his baubles
> is learning the action of light,
> motion, gravity, muscular force;
> and the game of human life,
> love, fear, justice, appetite,
> man, and God interact.
>
> —Ralph Waldo Emerson (1802-1882)

Similarly, because children reach developmental milestones at different ages, movements should be viewed and selected using *developmentally appropriate* standards, not necessarily using an *age-appropriate* designation. The average age for a baby's first walking steps, for example, is 11.7 months. The age of walking onset ranges, however, from as early as 9 months to as late as 17 months. Clearly, not all infants and children reach developmental movement milestones at the same age.

Developmental factors include genetic, environmental, maturational, and cultural influences, none of which alone determines the course of development. Many factors combine to determine the course of motor development and the eventual movement capacity of each individual.

In Others' Words: Charles Darwin

Under a transport of joy or of vivid pleasure, there is a strong tendency to various purposeless movements, and to the utterance of various sounds. We see this in young children, in their loud laughter, clapping of hands, and jumping for joy; in the bounding and barking of a dog when going out to walk with his master; and in the frisking of a horse when turned out into an open field. Joy quickens the circulation, and this stimulates the brain, which reacts again on the whole body.

. . . it is chiefly the anticipation of a pleasure and not its actual enjoyment, which leads to purposeless and extravagant movements of the body, and to the utterance of various sounds. We see this in children when they expect any great pleasure or treat.

Moreover, the mere exertion of the muscles after long rest or confinement is in itself a pleasure, as we ourselves feel, as we see in the play of young animals. Therefore on this latter principle alone we might expect, that vivid pleasure would be apt to show itself conversely in muscular movements. (Darwin, 1998, pp. 80-81)

—*Charles Darwin*, The Expression of the Emotions in Man and Animals

Movement Considerations in Older Persons

Movement capability declines with age. Older people tend to move more slowly, with more limited range of motion, and with greater hesitancy than do younger, healthy adults. Older persons are more likely to be afflicted with certain diseases (e.g., heart disease, diabetes) and suffer from injuries (e.g., hip fractures). Other factors, such as slower reaction times, reduced muscle strength and balance, and fear of falling may also contribute to slowed movements. Subsequent chapters examine many of these factors in detail.

Undeniably, some slowing happens as we age. But do we slow down because we age, or do we age because we slow down? To some extent, the latter is true, largely because of the adoption of sedentary lifestyles. The movement capacity of older people who get little if any regular physical activity typically declines, often markedly. These individuals may deteriorate to the point where simple activities of daily living (ADLs) become a challenge. ADLs include tasks such as bowel and bladder control, personal grooming, toilet use, feeding, bathing, mobility, transfer, and climbing stairs. Taken together, ADLs may be used as an index of functional independence. A low cumulative index score may indicate a loss of independence and a lowered quality of life. That's the bad news.

The good news is that research clearly shows that the body can respond to training throughout the life span and that it is never too late to begin restorative training programs aimed at improving cardiorespiratory function, muscle strength, and balance. In a now-classic study, Fiatarone et al. (1990) noted strength gains of 174% after just 8 weeks of high-intensity resistance training in frail, institutionalized persons averaging 90 years. The researchers demonstrated that high-resistance weight training promoted gains in muscle strength, size, and functional mobility in persons up to 96 years of age!

Although some decline in function may be inevitable, with proper lifestyle choices—and a little luck—most of us have the potential to maintain productive movement well into our later years.

> You start chasing a ball and your brain immediately commands your body to "Run forward! Bend! Scoop up the ball! Peg it to the infield!" Then your body says, "Who, me?"
>
> —Joe DiMaggio, Baseball Hall of Famer (1914-1999), commenting on getting older

Differences in Movement Ability

Each of us demonstrates movement abilities falling somewhere along what can be termed an ability continuum. An elite athlete may exhibit high ability in several movement forms, such as the ability of a professional basketball player to jump high, run fast, and control his body while in the air. A professional dancer may show exceptional ability to control her arm and leg movements and maintain balance throughout a performance. At the other end of the ability continuum, a person with physical disabilities may have difficulty walking or controlling limb movements.

There are many different movement forms, including gross motor skills and fine motor skills. **Gross motor skills** involve moving and controlling the limbs, as when someone walks, runs, or jumps. **Fine motor skills**, in contrast, involve small movements (e.g., manipulating objects with the fingers), such as sewing, writing, or typing. Good manual dexterity (i.e., ability to move the fingers) is essential for many occupations and activities.

Some activities require a combination of gross and fine motor skills for successful performance. A wide receiver in American football, for example, needs good gross motor skills to run down the field and elude defenders but must also have sufficient fine motor skills and eye–hand coordination to catch the ball thrown by the quarterback.

Few people have high abilities across all movement forms. Someone with good gross motor skills, such as a soccer player, may have difficulty with fine motor tasks. In contrast, a concert pianist with exceptional fine motor skills may be much less capable of performing gross movements.

Many factors determine motor abilities, including natural genetic potential, practice, level of physical conditioning, motivation, disease, and injury. A gifted athlete, for example, may enjoy the good fortune of having athletic parents who passed favorable movement potential on to their child. That movement potential may be fully realized through proper training and

"I move, therefore I am"

Western civilization, at a fundamental level, values the mind over the body. The foundations of this philosophy can be traced in large part to the work of French philosopher René Descartes (1596-1650), who in formulating his philosophy dismissed any information that might be subject to error, such as information gathered from his five senses. In his *Discourse on Method*, Descartes notes that "And noticing that this truth—*I think, therefore I am*—was so firm and so assured that all the most extravagant suppositions of the skeptics were incapable of shaking it, I judged that I could accept it without scruple as the first principle of the philosophy I was seeking" (Descartes, 1998, p. 18). Thus he established for himself, and much of civilization to follow, that the mind was primal and the body was both separate and secondary. According to Descartes, "the soul through which I am what I am, is entirely distinct from the body" (p. 19).

The belief that the mind is more important than the body, and that the purest way to know is through intellectual means, persists and helps explain why study of human movement has been regarded by many as an area unworthy of scientific inquiry. One consequence of this philosophical approach is that many believe little can be learned from experiencing human movement. Those involved in exploring human movement would promote a contrary opinion, that there is much to be learned by both *studying* and *experiencing* movement in all its forms.

While Descartes believed in the primacy of the mind, he nonetheless appreciated the complexity of the human organism. In his *Discourse on Method*, he wrote, "For they will regard this body as a machine which, having been made by the hands of God, is incomparably better ordered and has within itself movement far more wondrous than any of those that can be invented by men" (p. 31).

There can be no doubt that movement is central to our existence. Our philosophy, therefore, must include the following truth: *I move, therefore I am.*

---

conditioning, or it may never be fulfilled because of lack of motivation or debilitating injury. A person not as genetically blessed may nonetheless excel through hard work, perseverance, and strong motivation.

Dysfunctional, or disabled, movement refers to limited or compromised ability to move. Many factors can cause movement disabilities, including genetics (e.g., an infant born with a congenital deformity), disease (e.g., shaking limbs seen in persons with Parkinson's disease), or injury (e.g., a fractured bone). Other factors may also lead to movement dysfunction. In describing pathological walking, Perry (1992) identifies four functional factors that can impair gait: deformity, muscle weakness, impaired control, and pain. A functional deformity such as excessive tightness in a muscle, tendon, or ligament can limit joint range of motion. Muscle weakness due to paralysis or muscular disease can compromise the muscle's ability to move the arms or legs. Impaired control refers to an impairment in the nervous system's ability to provide sensory input and tell the muscles what to do. Finally, pain can discourage a person from performing movement throughout a full range of motion—or from moving at all.

Anatomical Concepts

The study of dynamic human anatomy involves many details and considerable memorization. Although knowing the details is important, it is equally important not to lose sight of the forest for the trees. Keeping the larger context in mind is essential, and consideration of several anatomical concepts may provide a useful backdrop for the study of the details. These concepts include anatomical complexity, variability, individuality, adaptability, connectivity, and asymmetry. Each of these concepts is discussed briefly in this section.

Complexity and Variability

We are complex organisms. The human body is composed of more than 200 bones and 600 muscles, hundreds of tendons and ligaments, miles of circulatory vessels, and countless nerve cells. These tissues are connected to one another and interact in many and wonderful ways. We have learned much about the human body's structure and function, but much more remains a mystery. Thousands upon thousands of researchers spend their professional lives trying to unravel these mysteries and enhance our knowledge of the human organism.

Humans come in all shapes and sizes. The basic components, for the most part, are the same in each of us. With few exceptions, each of us has the same number of bones, muscles, and internal organs. However, the size, shape, composition, and function of these components can vary greatly. At a gross level, some individuals exceed 7 feet in height (over 2 meters), while others are less than 4 feet tall (just over 1 meter). Body weights vary greatly. Skin color spans the spectrum from light to dark. Some individuals have large, strong, and dense bones. Others have fragile bones that may easily fracture. The chapters to follow present the basics of anatomy and function, but it is important to keep in mind that because we are not all built the same, we will not function alike. Certain structural characteristics may favor one individual over another when it comes to movement proficiency and performance capability.

The importance of anatomical variation cannot be overstated. As noted in the preface of the *Compendium of Human Anatomic Variation*, "Most modern textbooks of anatomy are more or less devoid of information on variations that so commonly appear in the dissecting room and, more importantly, in practice. Variations must be considered to be normal and thus must be anticipated and understood. It has been repeatedly stated in the literature that textbook descriptions are accurate… in only about 50-70% of individuals. From the standpoint of utilization of anatomic information in a clinical setting, textbooks are not only inadequate but may be dangerously misleading as well" (Bergman, Thompson, Afifi, & Saadeh, 1988, p. v).

Individuality and Adaptability

Another important anatomical concept closely related to structural variability is individuality. Each of us has individual characteristics and circumstances that affect our anatomy and physiology. These characteristics include gender, age, environment, genetics, ethnicity, cultural background, and family and individual history.

But as individual as we are, collectively we have a remarkable ability to adapt. Human body adaptations can be physiological or structural; they can be advantageous or detrimental. Muscles, for example, can increase in size (**hypertrophy**) and strength in response to resistance training or decrease in size (**atrophy**) when disused as a result of disease, injury, or a sedentary lifestyle. The cardiovascular system can improve its ability to transport oxygen following endurance training or have its functional capacity compromised by physical inactivity. Tissues such as bone can structurally adapt to the forces imposed on them over time. The nervous system (brain, spinal cord, nerves) has the capacity to modify its structure and function as it learns or is injured. This ability to adapt to environmental stimuli and events is termed **plasticity.**

All the body's tissues and systems can adapt to imposed demands. In general, there is a level of demand, or stimulus, that allows for optimal adaptation. A stimulus that is too small (e.g., from physical inactivity) reduces the body's performance ability and structural integrity, while one that is too great can result in tissue injury, such as when bones are overloaded and break.

Connectivity

Traditional approaches to studying anatomy logically begin with examination of the individual structural components (e.g., bones, muscles, tendons). Although this is an appropriate starting place, these tissues are connected to one another in the body, and the coordinated function of the whole organism depends on the connective characteristics. For example, muscle and tendon are joined at the **musculotendinous** (also **myotendinous) junction.** Here the muscle fibers intertwine with collagen fibers of the tendon in a structural arrangement designed to enhance surface area for attachment and to strengthen the site.

Forces applied to the body are transmitted from one tissue to another through their connections, and structural units (e.g., bone–ligament–bone) frequently fail at the attachment sites. Thus, we need to keep in mind the connectivity of the body's structure and the important role it plays in effective performance, tissue adaptation, and injury prevention.

Asymmetry

At first glance, the human body appears to be designed symmetrically: two each of arms, legs, eyes, ears, and so on. From the outside, our left and right sides, generally speaking, appear to be mirror images of each other. Further exploration, however, reveals that in many ways we are not designed symmetrically. The heart is not located along the body's midline, as symmetry would dictate. Similarly, the liver is on the right side, the spleen on the left. The right lung has three sections (lobes), whereas the left lung has two.

Joints in the body also are not symmetrical. The knee joint, for example, has a number of structural asymmetries. The supporting ligaments on the sides of the joint (collateral ligaments) are arranged differently. The ligament on the inner (medial) side of the knee structurally blends with the joint capsule that surrounds the joint. The ligament on the outer (lateral) side of the joint is structurally distinct and is located outside the capsule. The size and shape of the bony contact surfaces on the medial and lateral sides are not the same; nor are the cartilage discs (menisci) that help the knee function properly.

Anatomical asymmetry has allowed the body to develop into the complex organism it is, with all of the necessary organs neatly but asymmetrically packaged inside. Finding the genes responsible for determining what goes where is but one of the many anatomical challenges facing developmental geneticists. Structural asymmetries often have functional advantages, and it is important to keep this anatomical concept in mind as we learn more about the details of human anatomy and movement.

Levels of Structural Organization

Early study of the human body and its structure was limited to what could be observed by the unaided eye. Only when microscopic techniques were developed could the body's smaller structures be investigated. We now know that the larger (macroscopic) structures are composed of smaller structures, which in turn are made of even smaller structures. To view the spectrum of structural complexity, we describe the body's levels of structural organization.

The most basic organizational level is the chemical, or molecular, level. The human body is composed of more than a dozen elements, but four of these elements (hydrogen, oxygen, nitrogen, carbon) account for nearly all (>99%) of the atoms in the body. Atoms can be combined to form small molecules, such as water (H_2O) and oxygen (O_2), or larger molecules (macromolecules). The molecular composition of the body consists of water (67%) and macromolecules categorized as proteins (20%), lipids (10%), and carbohydrates (3%).

The next higher organizational level consists of cells, the smallest living units in the body. The human body contains trillions of cells that come in many different shapes and sizes. Specific cell types develop from undifferentiated cells known as **mesenchymal cells**, or **stem cells**.

At a higher level, cells combine to form tissues. **Tissues** are composed of similar cells (with specialized functions) and surrounding noncellular material that gives each tissue its structure and determines, in large part, its mechanical properties. The study of tissue structure is known as **histology**. There are four primary tissue types: epithelial, muscle, nerve, and connective.

It is a gratification to me to know that I am ignorant of art, and ignorant also of surgery. Because people who understand art find nothing in pictures but blemishes, and surgeons and anatomists see no beautiful women in all their lives, but only a ghastly stack of bones with Latin names to them, and a network of nerves and muscles and tissues.

—Mark Twain (1835-1910)

- *Epithelial tissue* is found throughout the body. It covers and lines body surfaces both inside and out, provides protection for adjacent tissues and organs, and helps regulate secretion and absorption.
- *Muscle tissue,* unlike any other tissue in the body, has the unique ability to produce force. The body uses these forces, for example, to move the arms and legs, pump blood, and facilitate transport of materials in the digestive and cardiovascular systems.
- *Nerve tissue,* found in the brain, spinal cord, and peripheral nerves, acts as the body's communication system. It receives sensory signals (e.g., pain, heat), integrates this sensory information at various levels of the brain and spinal cord, and then sends instructions to muscles and glands throughout the body.
- *Connective tissue* serves multiple functions. In general, connective tissues connect, bind, support, and protect. The noncellular matrix of connective tissues is especially important because it determines each connective tissue's mechanical properties.

Organs are structures composed of two or more tissues that have a definite form and function. The heart, for example, is an organ containing muscle tissue (to pump blood), nervous tissue (to control electrical activity), connective tissue (to hold the tissues together), and epithelial tissue (to protect the heart). An **organ system** consists of a group of organs that work together to perform specific functions. There are 11 organ systems:

Integumentary

Skeletal

Muscular

Nervous

Endocrine

Cardiovascular

Lymphatic

Respiratory

Digestive

Urinary

Reproductive

The highest organizational level is the organismic level, which combines all of the organ systems to form an *organism,* or a single living individual. Organisms are distinct from nonliving things in their ability to perform certain physiological processes. These include metabolism (i.e., the chemical processes that occur in our bodies), responsiveness (e.g., to stimuli such as pain), growth and differentiation, reproduction, and movement.

Muscle Tissue

Just as an automobile contains an engine to make it go, the human body has its own engines that provide the power needed to allow us to move. These engines are our muscles. They have a unique ability among all of the body's tissues to generate force and to contract. Their specialized cells allow muscle to produce *tensile* forces (sometimes simply referred to as *tension)* and to change their shape by shortening, or contracting.

The neural, biochemical, and mechanical details of how our muscles generate force are quite elegant and complex and will be considered in detail in chapter 4. Here we consider some of the basic characteristics of muscle tissue.

Muscle tissue is classified as one of three types: cardiac, smooth, or skeletal. Some of the structural distinctions between the three muscle types are shown in figure 1.3.

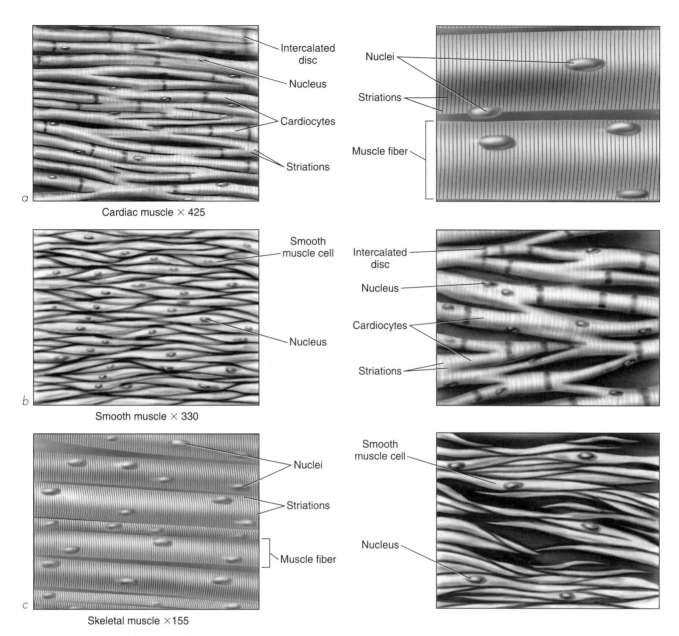

Figure 1.3 Magnified views of (a) skeletal muscle, (b) cardiac muscle, and (c) smooth muscle.

Cardiac muscle tissue is located only in the heart and is responsible for generating the forces required for the heart to pump blood out to the lungs and the rest of the body. The cardiac cells are tightly connected to one another and function, in normal situations, under the coordinated control of pacemaker cells. When the pacemaker cells are dysfunctional, the cardiac cells act randomly in a pattern known as *fibrillation*. If an external defibrillator is not used to shock the cells back into a coordinated pattern of contraction, death may result.

Smooth muscle is found in and around structures in the circulatory, respiratory, digestive, urinary, and reproductive systems; it facilitates movement of substances through tracts in those systems.

From a movement perspective, the most important muscle type is **skeletal muscle**. The human body contains more than 600 skeletal muscles, but many of these muscles are relatively small and play little or no role in human movement. Fewer than 100 pairs of skeletal muscles account for the vast majority of movement production and control. The most important of these muscles are discussed in chapter 5.

Connective Tissue

Connective tissue, the most abundant tissue type in the body, encompasses an array of individual tissues with vast differences in structure, function, and mechanical characteristics. Collectively, connective tissues provide support and protection, serve as a structural framework, and help bind tissues together. They also fill the space between cells, tissues, and organs; produce blood cells; store fat; protect against infections; and help repair damaged tissues. No single connective tissue, however, performs all these functions.

As with all tissues, connective tissues contain cellular and noncellular components. The cells are primarily responsible for each tissue's physiological functions. Connective tissues contain specialized cells unique to the tissue's specific functional requirements.

The noncellular portion (also termed **extracellular matrix** or **intercellular substance**) largely determines the mechanical characteristics of each connective tissue. The noncellular mineral component of bone, for example, gives bone its hardness and resistance to fracture. Connective tissues are distinguished by a relative abundance of extracellular matrix as compared with other tissues. The matrix is composed of fibers and ground substance.

The extracellular **ground substance** is the primary determinant of whether a connective tissue is solid (e.g., bone), fluid (e.g., blood), or somewhere in between (e.g., tendons, ligaments). In bone, the ground substance is a combination of primarily calcium phosphate and, to a lesser extent, calcium carbonate. The rigidity of these calcium salts is complemented by the flexibility of collagen fibers in bone's matrix to produce a structure that combines the best of both worlds: a tissue that is fracture resistant and remarkably well designed to meet the demands of everyday living. At the other end of the mechanical spectrum is blood, whose ground substance is its *plasma*. Blood contains cells (red blood cells, white blood cells, platelets) surrounded by a liquid (plasma) that facilitates the transport of materials through the cardiovascular system.

We now provide some details of several connective tissues (tendons, ligaments, cartilage) that play an important role in the anatomy of human movement. A detailed examination of bone is presented in the next chapter.

Tendons

A **tendon** is a cordlike connective tissue that connects skeletal muscle to bone (figure 1.4). The cellular component of tendons consists of many fibrocytes embedded in an extracellular matrix composed primarily of collagen fibers. The function of the tendon is to transmit the forces generated by the muscle to the bone to produce and control movement. To do this, the tendon's collagen fibers are aligned parallel to the tendon's line of action. The body also uses broad tendonlike sheets known as **aponeuroses** to cover a muscle's surface or to connect the muscle to another muscle or other structures.

The connective structure of a tendon creates three structural zones: the body of the tendon, the connective region between tendon and bone (**osteotendinous junction**), and the connective region between tendon and muscle (**myotendinous junction**, or **musculotendinous junction**). The osteotendinous junction contains microscopic transition zones of progressively stiffer tissues between the tendon and the bone. This structural feature distributes the load being transmitted from the muscle to the bone, thus reducing the chance of injury.

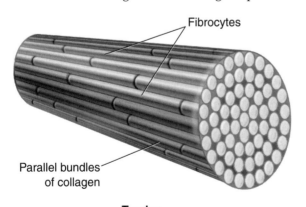

Fibrocytes

Parallel bundles of collagen

Tendon

Figure 1.4 Tendon structure. Note the parallel arrangement of collagen fibers.

The myotendinous junction similarly attempts, through its structure, to distribute forces passing through it. The collagen fibers of the tendon are intertwined with folded cell membranes of the muscle.

In normal situations, the tendon performs its force-transmission function well. If the forces being transmitted exceed the strength of the fibers (or the whole tendon), however, the tendon may be injured. Chronic overuse of tendon may result in an inflammatory response, or **tendinitis.** This inflammation may affect the tendon itself or related structures such as the tendon's outer covering (sheath) or associated **bursa** (fluid-filled sacs that help cushion or reduce friction) adjacent to the tendon.

Ligaments

A **ligament** is a connective tissue that joins bone to bone, and its primary function, like that of tendons, is to resist tensile forces (figure 1.5). Ligaments protect bone-to-bone connections by resisting excessive movements or dislocation of the bones. In that role, ligaments are sometimes referred to as *passive joint stabilizers.* A ligament's fibers (primarily collagen with some elastic and reticular fibers) are oriented according to their function. The fibers may be arranged parallel, obliquely, or even in a spiral formation. In addition, some ligaments are easily identifiable, isolated structures. Others appear as indistinct thickenings of the fibrous joint capsule that surrounds some joints.

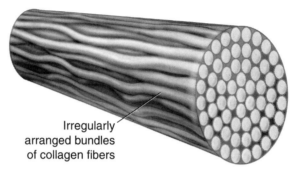

Irregularly arranged bundles of collagen fibers

Ligament

Figure 1.5 Ligament structure. Note the irregular arrangement of collagen fibers.

Joint ligaments contain a variety of sensory receptors capable of providing the nervous system with information about movement, position, and pain. The exact role of these sensory structures remains somewhat controversial and is the subject of ongoing study.

Ligament injury (**sprain**) occurs when applied forces exceed the tensile strength of the ligament's extracellular matrix. The severity of injury can range from mild to severe. Mild sprains are characterized by local tenderness, no visible injury, and no loss of joint stability. Moderate sprains involve a partial tearing of the ligament, visible swelling, marked tenderness, and some loss of stability. Severe sprains result in complete tearing of the ligament with gross swelling, considerable pain, and joint instability.

Cartilage

Cartilage is a stiff connective tissue whose ground substance is nearly solid. Cartilage cells (chondrocytes) have the ability to produce both the fibers and ground substance of the extracellular matrix. The types and amount of matrix constituents distinguish three kinds of cartilage: hyaline cartilage, fibrocartilage, and elastic cartilage. None of these cartilage types has intrinsic blood, lymph, or nerve supply. The absence of circulatory structures makes it necessary for cartilage to receive its nutrients and remove metabolic waste through diffusion.

Hyaline cartilage gets its name from its glassy appearance (*hyalos* = glass) and is the most common cartilage type. It is found on the articular surfaces of most joints, at the anterior portions of the ribs, and in components of the respiratory system (e.g., trachea, nose, bronchi). Hyaline cartilage also serves as the precursor to bone in the developing fetus.

Both strong and flexible, **fibrocartilage** is functionally well suited to the role it plays at stress points in the body where friction could be a problem. Fibrocartilage also serves as a "filler" material between hyaline cartilage and other connective tissues and is found near joints, ligaments, and tendons and in the intervertebral discs.

Elastic cartilage, as its name suggests, has a great deal of flexibility. Its matrix contains elastic and collagen fibers. The matrix of elastic ligaments appears more yellowish in color than hyaline or fibrocartilage does because of its higher proportion of elastic fibers. Elastic cartilage is found in the external ear and portions of the respiratory system.

Anatomical References and Terminology

Although knowing and understanding the structural details of the human body are essential, so too is communicating this knowledge with others in the field. To facilitate communication, scientific disciplines have a generally agreed upon system or set of terms (**nomenclature**). The anatomical nomenclature uses terms to identify and describe body regions, body positions, movement planes, axes of rotation, joint positions, and movements. These terms allow us to specify the location of the body or any of its segments at a point in time and to describe joint movements. This common nomenclature allows professionals in many areas of kinesiology and related fields to communicate.

Body Positions and Body Planes

The standard body reference position is referred to as the **anatomical position** (figure 1.6). The body is erect with the head facing forward and the arms hanging straight down, palms facing forward. Anatomical position may be "unnatural," but it nonetheless provides us with a suitable reference point. In this position, all joints are said to be in a **neutral position**. Joint positions or ranges of joint motion are measured starting from these reference positions.

In anatomical position, the human body can be sectioned into three mutually perpendicular flat surfaces, or *planes,* as shown in figure 1.7. Sectioning the body in this way is useful for describing movements of the body and its segments. The three principal planes are the sagittal (also median), frontal (also coronal), and transverse planes. Each plane is at right angles to the other two. Movement occurring parallel to a given plane has that as its *plane of action.*

Sagittal planes, in general, divide the body into left and right sections. If a particular sagittal plane is placed along the midline, dividing the body in half, it is termed a **midsagittal** or **median plane**. Any sagittal plane offset from the midline to one side or the other is called a **parasagittal plane**. Body movements are said to occur "in the sagittal plane" if they are made

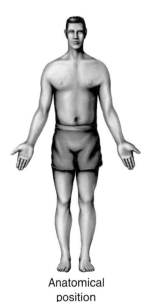

Anatomical position

Figure 1.6 Anatomical position.

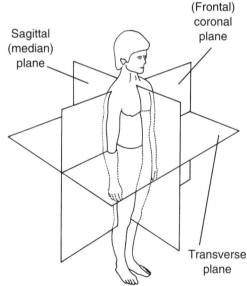

Figure 1.7 Principal body planes.

Reprinted, by permission, from J. Watkins, 1999, *Structure and function of the musculoskeletal system* (Champaign, IL: Human Kinetics), 55.

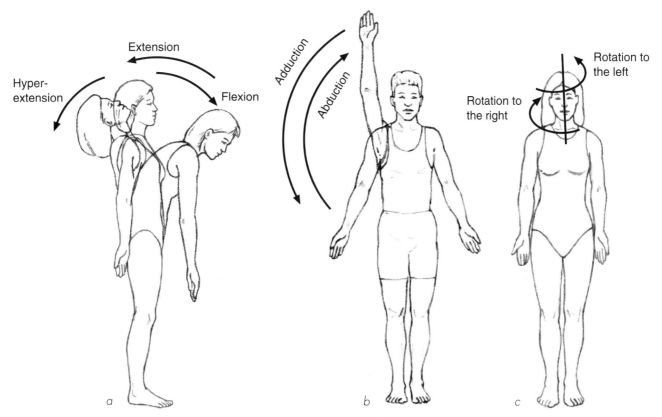

Figure 1.8 *Movement in each of the three primary planes: (a) sagittal, (b) frontal, and (c) transverse.*

Reprinted, by permission, from P. McGinnis, 2004, *Biomechanics of sport and exercise,* 2nd edition (Champaign, IL: Human Kinetics), 165.

parallel to the midsagittal plane. For example, forward bending of the trunk (figure 1.8a) and flexion of the elbow (from anatomical position) occur in the sagittal plane.

The **frontal plane** sections the body into anterior and posterior parts. Moving the arms away from the midline (figure 1.8b), leaning to one side or the other, and "jumping jacks" all occur in the frontal plane. The third principal plane, the **transverse plane**, divides the body into superior and inferior parts. Rotational movements such as twisting the head from side to side occur in the transverse plane (figure 1.8c).

Although it is helpful to define the three principal planes and describe movements in these planes, many human movements do not occur in a single plane and instead cross two or more. Description of these more complex movements often becomes problematic. If our movements were constrained to a single plane, they would appear "robotic" and unnatural. We present the planar description and analysis of movement as a starting point, with no presumption that such a scheme will be able to describe all movements.

Directional Terms and Axes of Rotation

Anatomical terms of direction describe the relative position, orientation, or direction of body surfaces, planes, or segments. Common terms are summarized in table 1.2. These terms typically occur in pairs (e.g., anterior and posterior), with each member of the pair having a meaning opposite to the other.

Rotational, or angular, movements of body segments happen around specific axes of rotation. An **axis of rotation** is defined as an imaginary line about which rotation occurs. The axis always is perpendicular (i.e., at a right angle) to the plane of action. For example, lifting (rotating) the arm away from the midline occurs in the frontal plane. The axis of rotation is a line through the center of the shoulder joint, with an anterior–posterior orientation perpendicular to the frontal plane. The axes of rotation for joint-specific movements are detailed in chapter 3.

Table 1.2 Anatomical Terms of Direction

Term	Description
Anterior (ventral)	In the front or toward the front
Anteroinferior	In front and below
Anterolateral	In front and to one side (usually outside)
Anteromedial	In front and toward the inner side or midline
Anterosuperior	In front and above
Bilateral	On two, or both, sides
Contralateral	On the opposite side
Deep	Away from the surface of the body
Distal	Farther away from the axial skeleton (trunk)
Inferior (caudal)	Toward the feet or ground, or below relative to another structure
Ipsilateral	On the same side
Lateral	Away from the midline of the body or structure, or farther from the midsagittal plane
Medial	Toward the midline of the body or structure, or closer to the midsagittal plane
Posterioinferior	Behind and below, in back and below
Posterior (dorsal)	Behind, in back, or toward the rear
Posterolateral	Behind and to one side (usually outside)
Posteromedial	Behind and toward the inner side or midline
Posterosuperior	Behind and above
Prone	Body lying on the belly with face downward
Proximal	Closer to the axial skeleton (trunk)
Superficial	Toward or nearer to the surface of the body
Superior (cephalic)	Toward the head, or above relative to another structure
Supine	Body lying on the back with face upward
Unilateral	On one side

The Study of Human Movement

The study of human movement requires a multidisciplinary perspective that considers psychological, physiological, anatomical, environmental, sociological, and mechanical factors (figure 1.9). Each of these areas offers unique insights into the complexities of human movement. Only by integrating information from all these areas, however, can we reach a comprehensive understanding and appreciation of human movement.

Anatomy, as described earlier, is the study of the body's structure and how the different structural components functionally relate to one another. Physiology deals with the functions of the body parts and systems and, in our current context, focuses on movement initiation and control. Biomechanics studies the mechanical aspects of movement and explores the role of force, time, and distance. Although our discussions in this book focus mostly on these three areas of study, movement is also greatly affected by psychological, sociological, and environmental variables.

Psychology deals with the mind and behavior. Psychological factors such as perception and motivation play an important role in how we move. An athlete in the throes of intense competition, for example, can move in ways that are not possible in a less-motivated state.

Sociology is the study of social institutions and social relationships. Sociological factors play an essential role in movement, such as in team settings when one person's performance can be enhanced or restricted by interactions with teammates and opponents.

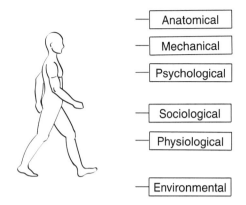

Figure 1.9 Multidimensional factors influencing human movement.

Human movement does not occur in isolation but rather is influenced by environmental factors. These external influences include weather, surface characteristics, apparel, and equipment. The walking pattern of a person struggling across a hot desert terrain would be much different from that of someone cautiously negotiating an icy sidewalk in the middle of winter.

We emphasize that each of these ways of studying human movement offers valuable information and insights, but only by taking a multidisciplinary approach can we truly understand movement. Before continuing with the next section, take a few minutes to consider the anatomical, physiological, mechanical, psychological, sociological, and environmental characteristics in each situational pair below. For each pair, list the similarities and differences you might expect to see.

- A 1-year-old baby taking her first hesitant steps on a carpeted floor compared with the walking pattern of an 80-year-old grandfather moving slowly with the assistance of a cane on a concrete sidewalk
- A soccer goalie trying to stop a ball rapidly approaching her goal compared with a long-distance runner struggling at the finish line of a 26.2-mile marathon
- A ballet dancer landing from a graceful leap compared with a snowboarder racing down a snowy mountain slope
- A middle-aged man maneuvering his wheelchair up a crowded street compared with a teenage girl dancing at her senior prom

concluding comments

The goal of this first chapter is to establish a conceptual framework and perspective of anatomy and human movement that will enhance our understanding of the human body and how it works. According to author John Jerome, in writing a treatise on the biophysical dynamics of movement and sport, "analyze us finely enough and in the end we are nothing more than electrochemical soup" (Jerome, 1980, p. 43). In one sense, Jerome is correct because we are a complex composite of chemical compounds. But in so many ways we are much more than just soup. Our bodies are composed of countless molecules arranged hierarchically to form cells, tissues, and organs. The organs are gathered in organ systems that collectively form the human organism. The integrated functioning of all the body's systems allows us to perform the myriad tasks essential to life, in particular the ability to move.

In the following chapters, you will receive the opportunity to go beyond mere memorization of details and grow in your understanding and appreciation of this most marvelous of creations, the human body. It is unparalleled in its complexity, variability, individuality, and adaptability. Keep in mind, too, that the human body is imperfect both in its structure and its function and subject to influences that limit movement proficiency.

critical thinking questions

1. Elaborate on the notion that movement is a fundamental property of life.
2. Define kinesiology, and explain the purpose and importance of studying kinesiology.
3. Motor behavior may be divided into three subdivisions: control, learning, and development. Describe each division, and explain how they contribute to our understanding of human movement.
4. Distinguish between developmentally appropriate standards and age-appropriate standards. Which one would you support and why?
5. Support and refute both sides of the following issue: Do we slow down because we age, or do we age because we slow down?
6. Compare and contrast gross versus fine motor skills. Give an example of when an athlete needs and uses both.
7. At the molecular level, what are the principal building blocks of the human body?
8. Describe the four primary tissue types, and give specific examples for each.
9. List the 11 organ systems that form the human body. Based on the information provided in this chapter, select one of the organ systems and describe its structure and function.
10. Describe the three planes of motion, and provide specific examples of movements we can perform in each plane.

suggested readings

Behnke, R.S. (2001). *Kinetic anatomy.* Champaign, IL: Human Kinetics.

Cappozzo, R., Marchetti, M., & Tosi, V. (1992). *Biolocomotion: A century of research using moving pictures.* Rome: Promograph.

Cech, D.J., & Martin, S. (2002). *Functional movement development across the life span* (2nd ed.). Philadelphia: Saunders.

Estes, S.G., & Mechikoff, R.A. (1999). *Knowing human movement.* Boston: Allyn & Bacon.

Hale, R.B., & Coyle, T. (1977). *Anatomy lessons from the great masters.* New York: Watson-Guptill.

Hale, R.B., & Coyle, T. (1988). *Albinus on anatomy.* Mineola, NY: Dover.

Jenkins, D.B. (1998). *Hollinshead's functional anatomy of the limbs and back* (7th ed.). Philadelphia: Saunders.

Keele, K.D. (1983). *Leonardo da Vinci's elements of the science of man.* New York: Academic Press.

MacKinnon, P., & Morris, J. (1994). *Oxford textbook of functional anatomy* (Vol. 1, rev. ed.). Oxford, UK: Oxford University Press.

Martini, F.H., Timmons, R.J., & McKinley, M.P. (2000). *Human anatomy* (3rd ed.). Upper Saddle River, NJ: Prentice Hall.

O'Malley, C.D., & Saunders, J.B. de C.M. (1997). *Leonardo da Vinci on the human body.* Avenel, NJ: Wings Books.

Saunders, J.B. de C.M., & O'Malley, C.D. (1973). *The illustrations from the works of Andreas Vesalius of Brussels.* Mineola, NY: Dover.

Schmidt, R.A. (1975). A schema theory of discrete motor skill learning. *Psychological Review, 82,* 225-260.

Thelen, E., & Smith, L.B. (2000). Dynamic systems theories. In W. Damon (Ed.), *Handbook of child psychology* (5th ed., pp. 563-634). New York: Wiley.

Thompson, C.W., & Floyd, R.T. (1994). *Manual of structural kinesiology* (12th ed.). St. Louis: Mosby.

Trew, M., & Everett, T. (2001). *Human movement: An introductory text.* Edinburgh: Churchill Livingstone.

two

Osteology and the Skeletal System

objectives

After studying this chapter, you will be able to do the following:

- Describe the functions of the skeletal system

- Name and describe the function of bone cells

- Explain the macroscopic and microscopic structure of bone

- Describe skeletal system organization, including bones of the lower body, upper body, and trunk

- Explain the processes of bone modeling and remodeling

- Explain the primary factors involved in bone health, including exercise, diet, and aging

As man has within himself bones as supports and protection of the flesh, so the world has rocks, supports of the earth.

—*Leonardo da Vinci (1452-1519)*

In light of the skeletal system's importance in supporting us, protecting us, and allowing us to move, it is both ironic and unfortunate that the image of the skeleton is often associated with fear and death. From the skeletons that abound during the frightful Halloween season to the stark picture of skeletal remains strewn across a desert landscape in a Hollywood movie, bones often are used to portray a feeling of lifelessness and dread. In reality, bones are dynamic organs that elegantly blend form with function and serve many useful purposes.

Functions of the Skeletal System

The skeletal system plays an integral role in mechanical and physiological functions essential for everyday life. Mechanically, the skeleton provides structural support for the body, acts as a lever system to permit movement of body segments and joints, and protects organs of the body's other systems. The bones of the skeletal system also act as a storehouse for minerals and contain tissues responsible for blood cell production. These five functions, varied as they are, all play critical roles in a comprehensive understanding of how the skeletal system facilitates human movement.

- *Structural support:* The skeletal system consists of 206 bones that collectively form a supporting framework and give shape to the body (figure 2.1). This framework bears applied loads with limited deformation, supports organs and soft tissues, and provides attachment sites for many muscles. Without bones, we might be structurally likened to an intelligent jellyfish. The bones also give shape, or form, to the body. Bones of the skull, for example, determine the shape of the head. The size and shape of bones often are related to the functions they perform. Women, for example, have a relatively wider pelvis than do men to allow for passage of a newborn during delivery. Similarly, the long bone (tibia) of the lower leg is shaped like a column to support the body's weight.

- *Movement:* Bones function as mechanical levers. When muscles (attached through tendons) pull on a bone, they change the magnitude and direction of forces and create rotational effects at joints (e.g., elbow, knee). The joints then rotate about an imaginary line known as the axis of rotation. Joint movements can range from the subtle motion of a violinist's fingers to complex, multijoint movements such as a vertical jump in which simultaneous extension of the hip, knee, and ankle powerfully propel the body upward.

- *Protection:* The hardness of bone makes it an ideal material to protect other more vulnerable organs in the body. For example, the skull protects the brain and eyes; the vertebrae shelter the spinal cord; the ribs protect the heart, lungs, and other visceral organs; and the pelvis shields organs of the urinary and reproductive systems. Without the protection provided by the skeletal system, the human body would be dramatically more susceptible to injury.

- *Mineral storehouse:* The calcium salts found in bone provide an important mineral reserve used by the body to maintain normal concentrations of calcium and phosphorus ions in body fluids. The body contains approximately 1 to 2 kg of calcium, 98% of which is contained in the bones in the form of calcium phosphate crystals. The small amount of remaining calcium is distributed throughout the body and is integral to physiological processes such as muscular contraction and nerve conduction.

- *Blood cell production:* The cavities and spaces in bone house a loose connective tissue known as **marrow**. There are two types of marrow, named according to their color. *Yellow marrow* contains large numbers of fat cells (**adipocytes**) that produce its characteristic color. The yellow marrow commonly resides in the marrow cavity in the shaft of long bones and provides a reserve source of energy. *Red marrow* consists of a mixture of red and white blood cells and platelets, along with the precursors, or stem cells, that produce them. The stem cells form blood cells in a process called **hematopoiesis** (also *hemopoiesis*). Important sites of blood cell formation include the proximal ends of the femur and humerus, sternum, ribs, and vertebrae.

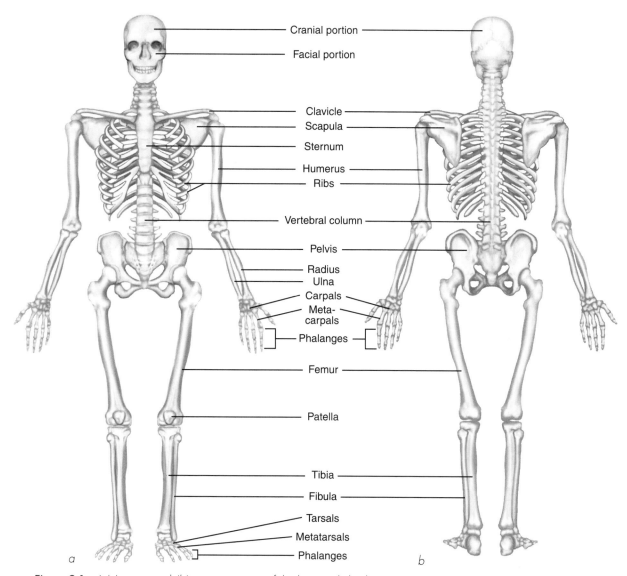

Figure 2.1   (a) Anterior and (b) posterior views of the human skeletal system.

Reprinted, by permission, from T.R. Baechle and R.W. Earle, *Essentials of strength training and conditioning*, 2nd edition (Champaign, IL: Human Kinetics), 27.

Bone Histology

As do all tissues, bone has a cellular component and a noncellular (extracellular) component. At the cellular level, specialized bone cells produce new bone and monitor and maintain their surrounding matrix. The extracellular component, or matrix, composed of mineral salt crystals and collagen fibers, gives bone its rigidity and strength. Because bones form the levers that allow us to move, healthy bones are essential for effective movement. Bone health relies on the action of bone cells and their surrounding matrix.

Bone Cells

The cellular component of bone is composed of four cell types, namely osteoprogenitor (mesenchymal) cells, osteoblasts, osteocytes, and osteoclasts. The **osteoprogenitor cells** are relatively few in number and are undifferentiated mesenchymal cells with the ability to produce daughter cells that can differentiate to become osteoblasts. The differentiation of mesenchymal cells into osteoblasts involves a 2- to 3-day process that seems to be triggered by mechanical stresses applied to the tissue.

The resulting **osteoblasts** are mononuclear (single nucleus) cells described as "producers" or "formers" because they produce the organic portion (**osteoid**) of the extracellular matrix. In performing this function, the osteoblasts are responsible for the materials required for new bone formation.

As an osteoblast matures, it becomes smaller and less metabolically active, finally becoming encapsulated by the calcified osteoid in a small pocket known as a **lacuna** (pl. *lacunae*). The osteoblast's role changes from forming new bone to monitoring and maintaining the bony matrix. These smaller mature cells are now called **osteocytes**.

Osteoclasts are large, multinucleated cells formed by the fusion of monocytes that originate in the red bone marrow. The osteoclasts are described as "resorbers" because their primary role is the **resorption**, or breakdown, of bone. These cells produce and release acids that demineralize the bone and enzymes that dissolve collagen fibers in the matrix.

Bone Matrix and Bone Structure

The extracellular matrix of bone consists of water, calcium phosphate (hydroxyapatite) crystals, collagen, and small amounts of ground substance. These elements are divided into organic and inorganic portions. The organic components (collagen and ground substance) are primarily produced by the osteoblasts and to a much lesser extent by the osteocytes. The collagen fibers account for about 30% of the bone's dry weight and provide resistance to tension, bending, and twisting.

The structure of bone can be considered at several levels. Human bones come in a wide variety of shapes and sizes and are classified at a gross level according to their shape: long, short, flat, irregular, or sesamoid (figure 2.2). Another bone classification is sutural, found in

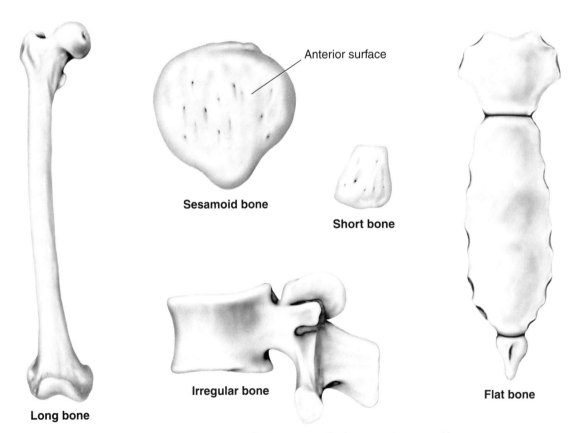

Anterior surface

Sesamoid bone

Short bone

Irregular bone

Flat bone

Long bone

Figure 2.2 Bone shape: long bone, short bone, flat bone, irregular bone, and sesamoid bone.

the skull. Bone shape plays an essential role in determining how much movement is allowed at a given joint, as will be explained in the next chapter.

Examination of bone's gross structure reveals two types of bone based on the density and structural arrangement of its components. At the microscopic level, each of these two bone types exhibits distinct structural characteristics.

Macroscopic Bone Structure

A cross section of long bone (figure 2.3) reveals two types of bone distinguished by the tissue's **porosity**. Bone porosity is measured by the amount of soft tissue. Bone with little soft tissue has low porosity, while bone with much soft tissue has high porosity. From the perspective of bone **density** (amount of hydroxyapatite crystals per unit volume), bones with high density have low porosity. Conversely, bones with low density have high porosity. In theory, bone porosity could range anywhere along a continuum from 0 to 100%. In reality, however, bones have either high or low porosity.

Bone with low porosity (5-10%) is called **compact** (also **cortical**) **bone**. Compact bone is found in the shaft (**diaphysis**) of long bones and forms the hard outer covering (**cortex**) of all bones. This protective outer bony surface layer sometimes is described as a **cortical shell**. The cortical shell is in turn covered by a fibrous connective tissue, known as the **periosteum**, that

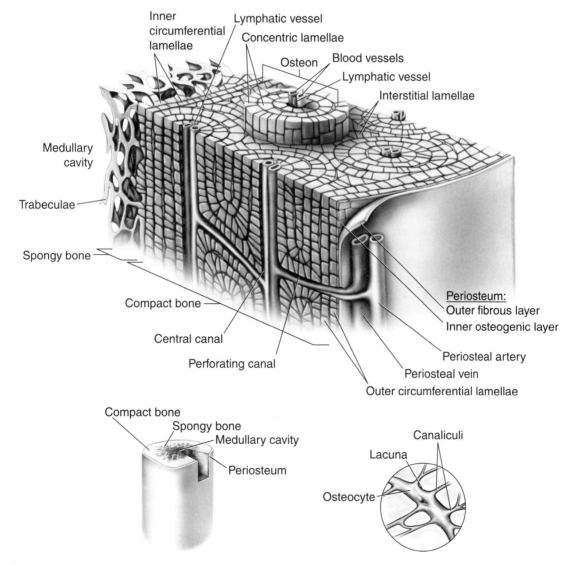

Figure 2.3 Cross-sectional structure of a long bone.

covers the entire bone except at the articular (joint) surfaces, which are covered by a protective layer of hyaline cartilage.

Bone with high porosity (75-95%) is termed **spongy** (also **trabecular** or **cancellous**) **bone**. Spongy bone is found under the cortical shell in vertebrae, flat bones, and the **epiphyses** of long bones. The bony matrix of spongy bone consists of plates of bone tissue known as trabeculae. Red bone marrow fills the space between the trabeculae.

It is important to emphasize that spongy and compact bone contain the same cellular and extracellular components; the difference is in the structure, not in the constituent materials.

Microscopic Bone Structure

Compact bone, when viewed by the unaided eye, appears as a solid, dense mass of bone. Microscopic evaluation, however, reveals a complex structure with a variety of connecting canals (figure 2.3). The fundamental structural unit of compact bone is the **osteon** (also the *Haversian system*). Each osteon consists of concentric layers (rings) of bone arranged around a central canal that houses blood vessels. The osteon's rings of bone, known as concentric lamellae, are configured like the rings around the bull's-eye of a target or the age rings in the cross section of a tree. Larger rings of bone, called circumferential lamellae, form the inner and outer boundaries of the bone's surface.

The capillaries of the central canal receive their blood supply from perforating canals (Volkmann's canals) that run transversely between adjacent osteons. These canals also connect osteons and facilitate inter-osteon nutrient supply and communication, as well as supply blood to the periosteum on the bone's outer surface.

In contrast to compact bone, spongy bone is much less organized and, at first glance, appears to be arranged randomly. Spongy bone is virtually devoid of osteons. However, the arrangement of spongy bone's bony plates (trabeculae) is far from random. The trabeculae are oriented to accept the forces applied to the bone.

Skeletal System Organization

The generally accepted number of bones in the skeletal system is 206. This number may, however, vary from person to person. Some individuals may lack certain bones, while others may have extra ones (e.g., small **sesamoid bones** that develop in tendons to reduce friction where tendons pass over bony prominences). The 206 bones are divided into two subsystems, or divisions: the **axial skeleton** and the **appendicular skeleton** (table 2.1). The primary purpose of the axial skeleton (skull, spinal column, and thoracic cage) is to protect and support internal organs and provide sites for muscle attachment. The appendicular skeleton, in contrast, consists of the bones and supporting structures of the upper and lower limbs and is involved in movements used in everyday life (e.g., walking, lifting).

Each of our bones has its own unique shape, size, and physical characteristics that match its function. We highlight here bones of the pelvic girdle, lower limbs, pectoral girdle, upper limbs, vertebral column, and thorax because these bones are the ones that support us and allow us to move from place to place and interact with our immediate environment.

Lower Body: Pelvic Girdle, Lower Limbs, and Arches of the Foot

 The **pelvic girdle** is similar to the pectoral girdle (which we will discuss in a moment) in that it connects limbs of the appendicular skeleton to the axial skeleton. However, distinct differences between the two girdles exist. The pelvic girdle, because of its role in weight bearing and locomotion, is much larger and heavier than the pectoral girdle, similar to the way that bones of the lower extremities are larger and stronger than their counterparts in the upper extremities. Collectively, the pelvic girdle, sacrum, and coccyx form the pelvis (figure 2.4).

Table 2.1   Bones of the Adult Skeletal System

Division of the skeleton	Structure	Number of bones
Axial skeleton	Skull	
	Cranium	8
	Face	14
	Hyoid	1
	Auditory ossicles	6
	Vertebral column	26
	Thorax	
	Sternum	1
	Ribs	24
		Subtotal = 80
Appendicular skeleton	Pectoral (shoulder) girdles	
	Clavicle	2
	Scapula	2
	Upper limbs (extremities)	
	Humerus	2
	Ulna	2
	Radius	2
	Carpals	16
	Metacarpals	10
	Phalanges	28
	Pelvic (hip) girdle	
	Hip, pelvic, or coxal bone	2
	Lower limbs (extremities)	
	Femur	2
	Fibula	2
	Tibia	2
	Patella	2
	Tarsals	14
	Metatarsals	10
	Phalanges	28
		Subtotal = 126
		Total = 206

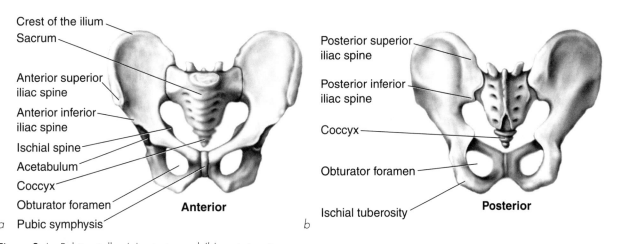

Figure 2.4   Pelvic girdle, (a) anterior and (b) posterior views.

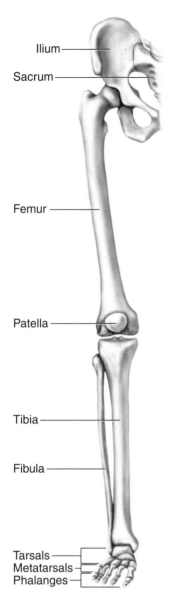

Ilium

Sacrum

Femur

Patella

Tibia

Fibula

Tarsals
Metatarsals
Phalanges

Figure 2.5 Bones of the lower extremity (anterior view).

Each lower extremity, or limb (figure 2.5), is made up of one bone (femur) in the thigh and two bones (tibia, fibula) in the lower leg (also shank). The foot contains 26 bones, including 7 tarsal bones, 5 metatarsal bones, and 14 phalanges (figure 2.6).

The bones of the foot are arranged to form two primary arches: the longitudinal arch, running from the calcaneus to the distal ends of the metatarsals, and the transverse arch, which extends from side to side across the foot (figure 2.7). The longitudinal arch is divided into a medial portion that includes the calcaneus, the talus, the navicular, three cuneiforms, and the three most medial metatarsals. The lateral portion is much flatter and is in contact with the ground during standing. The transverse arch is formed by the cuboid, cuneiforms, and bases of the metatarsals.

During weight bearing (e.g., walking, running, jumping), the arches compress to absorb and distribute the load. Several ligaments assist in this force distribution, including the plantar calcaneonavicular ligament ("spring" ligament), the short plantar ligament, and the long plantar ligament. The integrity of the arches and their ability to absorb loads are maintained by the tight-fitting articulations between foot bones, the action of intrinsic foot musculature, the strength of the plantar ligaments, and the plantar aponeuroses (plantar fascia).

Upper Body: Pectoral Girdle and Upper Limbs

 The **pectoral girdle**, or shoulder girdle, consists of two bones, the clavicle and scapula (figure 2.8). These two bones work in a coordinated fashion to facilitate the complex and necessary movements of the pectoral girdle and the arm. The most prominent joint of the pectoral girdle is the glenohumeral (shoulder) joint where the head of the upper arm bone (humerus) articulates with the scapula. The three-dimensional nature of the glenohumeral joint allows exceptional range of motion for activities such as throwing and lifting.

Each upper extremity, or limb (figure 2.9), is made up of one bone (humerus) in the brachium, or upper arm (sometimes called just the arm), and two bones (radius, ulna) in the antebrachium, or forearm. The wrist and hand contain 27 bones, including 8 carpal bones, 5 metacarpal bones, and 14 phalanges (figure 2.10). The bones and associated joints of the upper limbs provide for both large-scale (gross) movements such as pushing, pulling, and throwing and small-scale (fine) movements such as writing, gripping, and grasping.

Trunk: Vertebral Column, Spinal Curvatures, and Thorax

 The vertebral column (or spine) consists of 26 bones (vertebrae) that span from the base of the skull down to the pelvis (figure 2.11). The spine is divided into five regions: cervical region (7 vertebrae), thoracic region (12 vertebrae), lumbar region (5 vertebrae), sacrum, and coccyx. The sacrum is a single bone formed during development by the fusion of 5 sacral vertebrae. The coccyx is formed by the fusion of (usually) 4 vertebrae.

Each vertebra is identified by shorthand notation that specifies the spinal region and the number of the vertebra within that region. The most superior vertebra in each region is number 1. For example, the most superior vertebra in the cervical region is labeled C1. Underneath C1 is C2 and so on through C7. Similarly, the thoracic vertebrae are numbered T1 to T12, and the

Plantar view

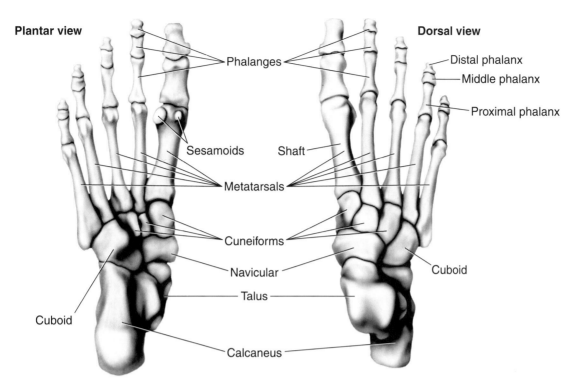

Phalanges

Sesamoids

Shaft

Metatarsals

Cuneiforms

Navicular

Talus

Cuboid

Calcaneus

Dorsal view

Distal phalanx

Middle phalanx

Proximal phalanx

Cuboid

Figure 2.6   Bones of the foot (plantar and dorsal views).

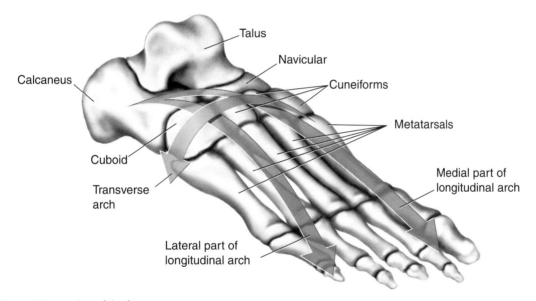

Talus

Navicular

Calcaneus

Cuneiforms

Metatarsals

Cuboid

Medial part of longitudinal arch

Transverse arch

Lateral part of longitudinal arch

Figure 2.7   Arches of the foot.

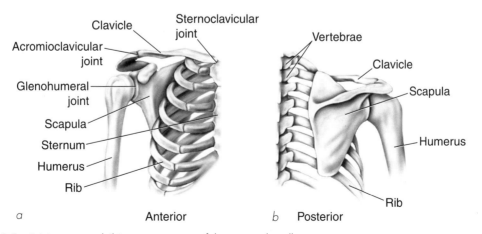

Clavicle

Sternoclavicular joint

Acromioclavicular joint

Glenohumeral joint

Scapula

Sternum

Humerus

Rib

Vertebrae

Clavicle

Scapula

Humerus

Rib

a   Anterior

b   Posterior

Figure 2.8   *(a)* Anterior and *(b)* posterior views of the pectoral girdle.

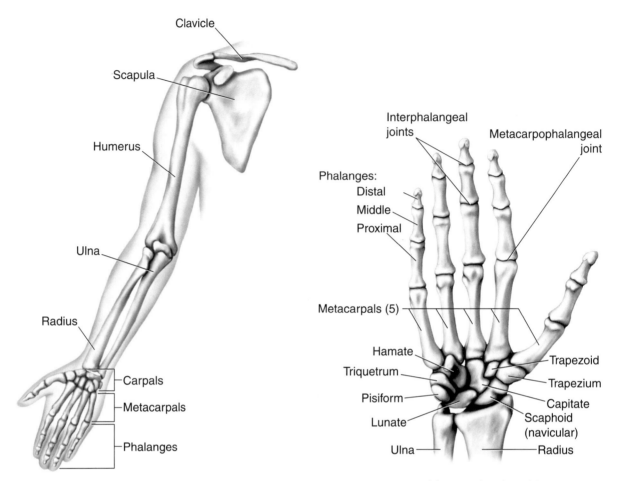

Figure 2.9 Bones of the upper extremity (anterior view).

Figure 2.10 Bones of the wrist, hand, and fingers.

lumbar vertebrae are designated as L1 to L5. The superior surface of the fused sacral bone is identified as S1.

Most vertebrae have a common structural plan, consisting of an anteriorly located vertebral body (or centrum), a vertebral arch (also neural arch) that extends posteriorly, and various processes that project from the vertebra and serve as muscle and ligament attachment sites or as articulation sites with adjacent vertebrae or ribs (figure 2.12a). The vertebral body is typically oval shaped (as viewed from above, see figure 2.12b) and relatively thick. One of its primary functions is to accept and transfer compressive forces acting through the spine. Moving caudally (inferiorly) from the head, each successive vertebral body must support increasing weight.

The vertebrae also are integrally involved in movements of the trunk and spine. Each vertebra moves relative to its adjacent ones; the summation of these movements determines the overall movement of the spine.

When viewed from the front or back, the normal spine is straight. From the side, however, the spine has characteristic curvatures (figure 2.11). In the fetus, there is a single anteriorly concave curvature. About 3 months after birth, when the infant begins to hold her head erect, a posteriorly concave curvature develops in the cervical region. Similarly, when the infant begins to walk, a curvature appears in the lumbar region. The thoracic and sacral regions retain their original curvatures, hence the term **primary curvatures**. Because they develop later, the oppositely directed curves of the cervical and lumbar regions are called **secondary curvatures**. The spinal curvatures serve important mechanical functions, including balance, shock absorption, and movement facilitation.

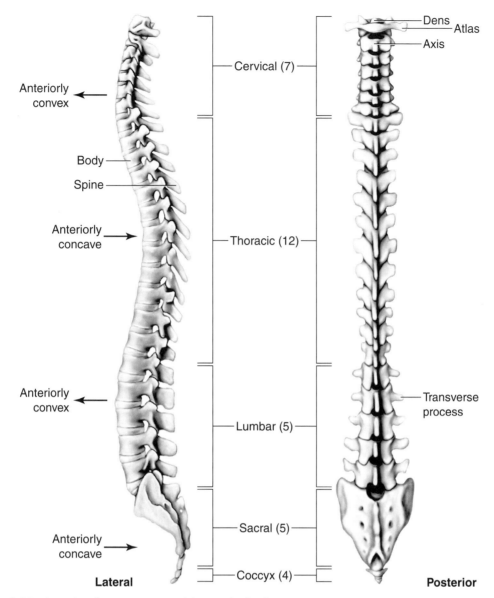

Cervical (7)

Anteriorly convex ←

Body

Spine

Anteriorly concave →

Thoracic (12)

Anteriorly convex ←

Lumbar (5)

Anteriorly concave →

Sacral (5)

Coccyx (4)

Lateral

Dens — Atlas
Axis

Transverse process

Posterior

Figure 2.11   Lateral and posterior views of the vertebral column.

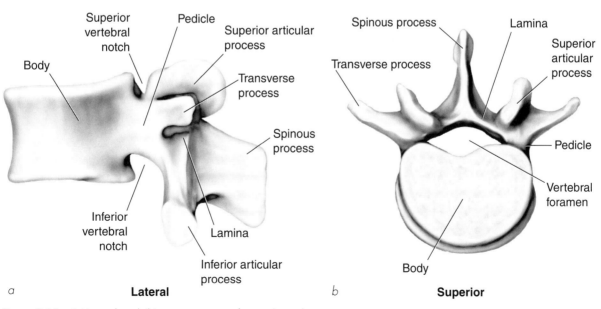

Superior vertebral notch

Pedicle

Superior articular process

Body

Transverse process

Spinous process

Inferior vertebral notch

Lamina

Inferior articular process

a   **Lateral**

Spinous process

Lamina

Transverse process

Superior articular process

Pedicle

Vertebral foramen

Body

b   **Superior**

Figure 2.12   (a) Lateral and (b) superior views of typical vertebra.

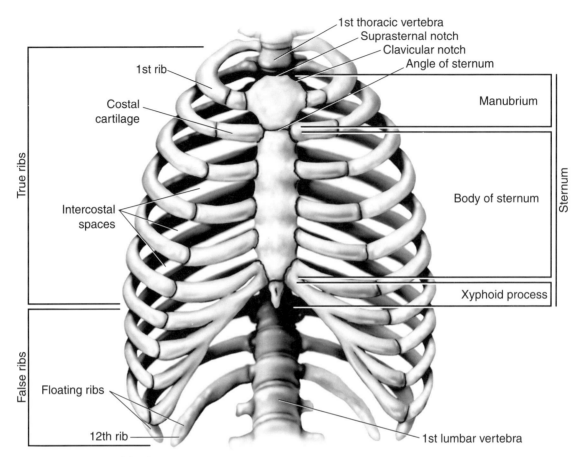

Figure 2.13   Bones of the thorax.

The sternum and ribs form skeletal components of the thorax (chest) (figure 2.13). The sternum (breastbone) is a centrally located flat bone made up of three sections (manubrium, body, xyphoid process). The manubrium and body join each other to form the sternal angle.

The manubrium also articulates with the proximal end of the clavicle and the first and second ribs. Ribs 2 to 10 attach directly or indirectly to the sternal body, with rib 2 also attaching to the manubrium.

The thorax contains 12 rib pairs that collectively encase and protect important internal organs such as the heart and lungs. The first 7 rib pairs articulate directly to the manubrium or sternal body and are thus called *true ribs*. The next 3 pairs attach indirectly through costal cartilage to the sternum. Because these ribs lack direct connection to the sternum, they are termed *false ribs*. The cartilage of ribs 8 to 10 (vertebrochondral ribs) merges with the cartilage of rib 7, which then connects with the sternum. The remaining two rib pairs (11-12) have no connection at all with the sternum and are therefore called *floating ribs*.

Bone Adaptation

Bone has an unfortunate reputation of being an inert, lifeless structure. Nothing could be further from the truth. Bone is a remarkably dynamic tissue that adapts to multiple internal factors (e.g., hormone levels, calcium concentrations) and external factors (e.g., mechanical loads). Adaptation has been defined as "modification of an organism or its parts that makes it more fit for existence under the conditions of its environment" (Mish, 1997, p. 13). This definition concisely characterizes the adaptability of bone and how it meets the changing needs imposed by activity levels and environmental demands.

Modeling refers to the addition (formation) of new bone. This process typically occurs during the growing years and is characterized by continuous bone deposition occurring on any bony surface that produces a net gain in bone. During modeling, osteoblasts and osteoclasts are not active along the same surface; resorption occurs along one surface while deposition occurs along another. For example, in the development of a long bone, osteoblasts continuously add bone to the periosteal (outer) surface, while osteoclasts enlarge the central canal by resorbing bone on the endosteal (inner) surface.

Remodeling is the resorption and replacement of existing bone. There is solid evidence that the remodeling of bone is, in fact, triggered by microdamage to the bone. In remodeling, the action of the osteoblasts and osteoclasts is tightly coupled, and the overall effect on bone is determined by the *net* activity of these cells. If osteoblast activity surpasses osteoclast activity, a net increase in bone mass results. Conversely, if osteoclast activity predominates, bone mass will decrease. This remodeling process is essential for maintaining the bone strength required to support the body during weight-bearing movements such as walking, running, and jumping.

Bone Health

Bone health is affected by numerous genetic, environmental, nutritional, hormonal, and mechanical factors. The complex interaction of these factors ultimately determines the physiological and biomechanical integrity of bone tissue. Bone research before the mid-20th century emphasized the physiological function of bone. In more recent decades, exploration of bone function and adaptation has widened to include other areas of scientific inquiry and clinical application. Although much has been learned about the requisites for bone health, much still remains a mystery. One thing that is *not* a mystery is the fact that bones need to experience forces to remain healthy. Human movements that involve weight-bearing forces are essential for the maintenance of bone health. Activities such as running and jumping, for example, subject bones of the lower extremities to high compressive forces. These forces stimulate the remodeling process and strengthen bones.

Exercise

There is little doubt that exercise and physical activity stimulate bone remodeling. The details of the phenomenon are extremely complex and involve the interaction of many variables, including exercise type and intensity, skeletal maturity, bone type, anatomical location, and hormone levels.

The primary relationships between physical activity and bone mass can be summarized as follows: (1) growing bone responds to low-to-moderate exercise through significant deposition of new cortical and trabecular bone; (2) bone responds negatively, by suppressing normal modeling activity, when certain activity thresholds are exceeded; (3) moderate-to-intense physical activity can elicit modest increases (1-3%) in bone mineral content in men and premenopausal women and may generate modest and site-specific increases in bone mass in postmenopausal women; (4) long-term benefits of exercise on bone health are maintained only by continuing exercise; and (5) bone mass gains appear to depend on initial bone mass (i.e., individuals with very low bone mass may show greater gains from exercise than those with only moderately reduced bone mass) (Whiting & Zernicke, 1998).

Exercise-related bone changes are usually specific to the region being affected (i.e., mechanically loaded) and are not systemic. For example, a person who runs on a regular basis would expect increased bone density in the legs but not in the arms. A softball pitcher would likely experience bone adaptation in her throwing-arm side but not in the other arm.

The dynamics of bone remodeling are important to keep in mind because the timing of the resorption–deposition cycle may leave the bone vulnerable to injury. If excessive loads are applied during the time between resorption and deposition, injury may occur. Athletes therefore are well advised to increase the intensity of their workouts gradually and progressively. Too much, too soon, and the athlete may end up on the sidelines tending to a fractured bone.

Too little exercise can also damage bone. Persons who choose a sedentary lifestyle or who are bedridden because of disease or injury are likely to lose bone mass and strength. Bone is deposited where it is needed and removed from where it is not. Thus, skeletal tissue not subjected to periodic mechanical loading from physical activity (e.g., walking or running) will atrophy. This net bone loss weakens the tissue and increases the chance of fracture.

One group at particular risk of bone loss is astronauts. Detectable bone loss has been measured after only a few days in space, and extended stays in space (i.e., weeks or months) can significantly compromise bone quality. Appropriate exercises while in space may reduce the negative effects of hypogravity on the astronauts' skeletal systems.

Diet

Nutrition profoundly affects bone health. Bone growth and remodeling are integrally dependent on having the proper kinds and quantities of nutrients required for building healthy bone. The body's normal bone mineral balance is typically well regulated by the synergistic actions of vitamin D, parathyroid hormone, and calcitonin. These substances control the absorption of dietary calcium, bone mineral deposition and resorption, and the renal (kidney) secretion and resorption of calcium and phosphorus. Adequate dietary calcium intake is essential because the body excretes calcium on a continual basis throughout the day. Calcium absorption is affected by vitamin D, dietary protein, phosphorus, fiber, and fat.

Insufficient levels of vitamin D may compromise calcium absorption and result in bones with reduced mineral content. These bones are softer and more "bendable" than healthy bones. Children with vitamin D deficiency may develop a condition called rickets, in which their more pliable bones, subjected to repeated forces, become bent. This may result in a bowlegged deformity because the bones lack the rigidity to withstand the forces of walking.

Adequate intake of vitamin C also is essential for bone health. Insufficient intake of this vitamin results in a disease called scurvy. In these cases, the lack of vitamin C compromises the synthesis of collagen. Low collagen levels result in brittle bones that may fracture easily.

Dietary protein levels affect the handling of calcium in the urine. Protein deficiency has been identified as a factor in the onset of bone mass reduction. Too little protein can lead to reduced calcium absorption in the intestine and increased urinary calcium levels. At the other extreme, too much dietary protein can lead to greater renal calcium loss and a negative calcium balance.

High levels of dietary fat and sugar have a negative effect on the absorption of calcium in the intestines. Therefore, low-to-moderate intake of fat and sugar is recommended for optimal bone health.

Aging

Many physiological, anatomical, and psychological changes accompany aging. Among these many changes are bone-specific alterations that affect the quality of bone tissue and consequently our ability to move effectively. During our growth period (from conception to approximately 25 years), the body normally increases its bone mass as osteoblast activity far exceeds the work of osteoclasts. This results in a net gain in bone as our skeletal system grows and develops. At some point, which varies across individuals and specific bones, we reach a level at which we have the greatest amount of bone we will ever have. This is referred to as **peak bone mass**. The level of peak bone mass is affected by many factors, including heredity, diet, and exercise.

Our middle years (25-50) are characterized by a relatively constant level or slight decrease in bone mass; osteoblasts and osteoclasts live in relative harmony in maintaining our skeletal integrity. As we move into our later years (50 plus), osteoblast activity decreases. This results in net bone loss and reduced skeletal strength. Postmenopausal women experience a more rapid rate of bone loss than do men, primarily because of reduced production of estrogen, a hormone that inhibits osteoclast activity.

One of the keys to bone health is accumulation of considerable bone mass in our early years of bone growth and development. If the inevitable decline in bone mass begins from a higher peak bone mass, the likelihood of eventually reaching a bone mass near the fracture threshold is greatly reduced.

What can we do to optimize bone mineral acquisition during the growing years? We can do many things, including making a lifelong commitment to weight-bearing exercise, engaging in a variety of vigorous short-duration activities, increasing muscle strength through resistance training, and avoiding immobility and prolonged sedentary periods. In addition, bone health can be enhanced by proper nutrition, adequate rest, limited alcohol intake, and not smoking. The benefits of these lifestyle choices are certainly not restricted to bone. All of the body's systems work best with a healthy combination of diet and exercise.

Osteopenia and Osteoporosis

Some degree of bone loss is a natural part of aging. Excessive bone loss, however, can have catastrophic consequences. **Osteopenia** refers to general bone loss. Serious bone loss with increased risk of fracture is termed **osteoporosis**. Osteoporotic bone exhibits excessive porosity and accompanying structural changes that greatly increase fracture risk. Spongy bone seems most prone to osteoporosis. All bones theoretically are at risk, but those with high levels of spongy bone (e.g., vertebrae and the proximal femur) are at particular risk.

Postmenopausal women have the highest risk of osteoporotic fracture. However, other populations also are at considerable risk for osteoporosis. Young female athletes, for example, with low body fat who train at high intensities commonly experience menstrual dysfunction. They may have no or very few menstrual cycles (**amenorrhea**) or irregular menstrual cycles (**oligomenorrhea**) and as a result have reduced estrogen levels. This creates an imbalance in the bone deposition–resorption ratio, which can eventually lead to osteoporosis. A comparison between amenorrheic elite athletes and a group with normal menstrual function found that the athletes with amenorrhea had up to 25% lower vertebral mineral mass than the control group (Marcus et al., 1985). Their problems are compounded by the fact that these athletes (e.g., runners, gymnasts) subject their bodies to excessive loading that further stresses the bones and increases the chance of fracture.

Osteoporosis and the Aging Population

Osteoporosis is a disease marked by reduced bone mineral mass and changes in bone geometry, leading to an increased probability of fractures, primarily of the hip, spine, and wrist. Progressive loss of bone mass can be a function of the normal aging process or can be caused by other disease processes. Many individuals are unaware of the existence of their osteoporosis, especially in its early stages.

Both men and women experience some loss of bone mass as part of normal aging, but osteoporosis progresses much more rapidly in postmenopausal women. In women, bone loss increases significantly for about 5 years after menopause and then slows to a more gradual loss.

Older individuals are at particular risk for falls and bone fractures. The statistics associated with the incidence of falls are sobering. For example, in the United States alone, more than 300,000 hip fractures happen annually. These statistics are particularly alarming in light of the fact that these injuries all too often are the first event in a chain that leads to eventual incapacity, dependence, and even death. Many of the characteristics associated with aging magnify fall risk. These include decreases in bone and muscle strength, impaired cardiorespiratory function, compromised visual acuity, slower reaction times, and reduced balance. The important issues related to the effects of aging are considered in greater detail in chapter 8.

Given that people now live longer, current demographic trends point to an exploding population of older people. This fact, coupled with more sedentary lifestyles, assures that the problems associated with osteoporosis will continue as a major health issue.

concluding comments

The skeletal system plays an important role in many processes needed for our survival and effective functioning, including structural support, protection, movement, blood cell production, and mineral storage. Healthy bones allow the skeletal system to perform all these functions well. Unhealthy bones, however, can dramatically affect our ability to perform even ordinary tasks of daily living. Many factors such as diet, hormones, activity level, and genetics interact to make bone health a multifactorial problem to which we have only some of the answers.

critical thinking questions

1. Describe the organic and inorganic components of bone. Be sure to include the functional significance of the four cell types common to all bones.
2. Describe the location and functional significance of red bone marrow.
3. Although compact (cortical) and spongy (trabecular or cancellous) bone contain the same cellular and extracellular components, explain how they differ in structure.
4. Discuss the beneficial effect of exercise on bone structure and remodeling.
5. Discuss how diet can affect bone integrity.
6. Briefly discuss some of the physiological and anatomical changes that occur in bones as we age. Include a brief discussion outlining the lifestyle choices we can make to offset the expected loss in bone integrity with age.
7. Describe osteoporosis, and briefly discuss why as we age it becomes more prevalent in women than in men.

suggested readings

Alexander, R.M. (1994). *Bones: The unity of form and function*. New York: Macmillan.

Benyus, J.M. (1997). *Biomimicry*. New York: Morrow.

Frankel, V.H., & Nordin, M. (2001). Biomechanics of bone. In M. Nordin & V.H. Frankel (Eds.), *Basic biomechanics of the musculoskeletal system* (3rd ed., pp. 26-55). Philadelphia: Lippincott Williams & Wilkins.

Martin, R.B., Burr, D.B., & Sharkey, N.A. (1998). *Skeletal tissue mechanics*. New York: Springer-Verlag.

three

Joint Motion and the Articular System

objectives

After studying this chapter, you will be able to do the following:

- Describe joint structure and classification

- Classify synovial joints according to their structure and function

- Explain the concepts of joint stability and mobility

- Describe movement planes and joint motion

- Describe types of joint movement

- Identify movements of the hip and pelvis, knee, ankle, and foot

- Identify movements of the shoulder, elbow, forearm, wrist, and hand

- Identify movements of the head, neck, and spine

- Describe spinal deformities (scoliosis, kyphosis, lordosis)

> Life is the ultimate example of complexity at work. . . . when [components] are combined into some larger functioning unit . . . utterly new and unpredictable properties emerge, including the ability to move.
>
> —Donald E. Ingber,
> The Architecture of Life

O ur ability to move depends on the body's joints. In this chapter, we present information on general joint structure and movement terminology, followed by details of the major joints involved in movement. Our approach emphasizes the functional aspects of joint structure, thereby providing you with the foundation needed to fully appreciate and understand the complexities of human movement in subsequent chapters.

Joint Structure and Classification

Our infinite capacity for movement depends on a musculoskeletal design that includes well-functioning articulations, or joints. Each joint's structure is well matched with its function. Some articulations allow considerable range of movement, or joint **mobility**. Others are built for joint **stability** and vigorously resist movement. The amount of joint mobility, or **range of motion**, is determined by both the structural congruity of the bones (i.e., how well they fit together) and the amount of support provided by tissues surrounding the joint (periarticular tissues). The shoulder (glenohumeral) joint, for example, has remarkable range of motion allowed by the relatively poor fit between the head of the humerus and the shallow glenoid fossa of the scapula. At the other extreme, suture joints of the skull are formed by bones that interlock in much the same way as pieces of a jigsaw puzzle fit tightly together and are thus immovable.

Joints can be classified in a number of ways. Most commonly, joints are *structurally* categorized according to the type of tissue that binds the joint together (binding tissue). Joints may also be classified *functionally* by the type or extent of movement they allow. Functional designations include joints that allow no movement (**synarthroses**), allow limited movement (**amphiarthroses**), or are freely movable (**diarthroses**). Logically, synarthroses and amphiarthroses tend to predominate in the axial skeleton, while diarthroses are more commonly found in the appendicular skeleton.

Structurally, joints fall into one of three categories based on the tissues that support the joint and bind the bones together and whether or not the joint contains a synovial cavity between the bones. **Fibrous joints** do not have a joint cavity and are bound by connective tissues composed primarily of collagen fibers. **Cartilaginous joints** also are void of a joint cavity but have cartilage as their binding tissue. The most common and complex articulations are **synovial joints**, which are distinguished by a fibrous joint capsule that surrounds and encapsulates the joint.

Fibrous Joints

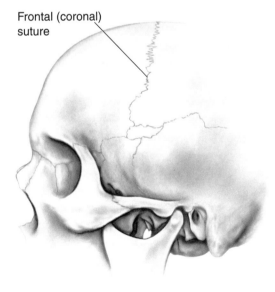

Frontal (coronal) suture

Figure 3.1   Suture joint.

The common features of fibrous joints include the absence of a synovial joint cavity and the presence of fibrous (collagenous) tissue to reinforce the bone junction. The two most common types of fibrous joints are sutures and syndesmoses. The amount of movement allowed at fibrous joints varies, but usually little or no movement is evident.

Suture joints connect bones of the skull. These interlocking bones are bound by a dense fibrous connective tissue that renders the joints immovable (figure 3.1). Suture joints therefore are functionally classified as synarthroses.

Syndesmoses are joints bound by ligaments, collagenous structures that connect bone to bone. The amount of joint motion may be limited, as in the articulation between the tibia and fibula, which are bound together by a collagenous **interosseous membrane**. The tibiofibular joint is functionally an amphiarthrosis. In contrast, considerable movement is allowed between the radius and ulna of the forearm. Here, the longer fibers of the interosseous membrane permit more extensive bone excursion and result in the radioulnar joint's functional classification as a diarthrosis (i.e., freely movable).

Cartilaginous Joints

Of the three types of cartilage, only hyaline cartilage and fibrocartilage are found as the binding tissue at cartilaginous joints. Elastic cartilage is never used to reinforce joints. Hyaline cartilage joints, or **synchondroses**, are found in the mature skeleton, for example, between the first rib and the manubrium (sternocostal joint) and temporarily in the developing skeleton between the diaphysis and epiphysis in the form of the **epiphyseal growth plate**, or **growth plate**.

Symphyses are joints between bones separated by a fibrocartilage pad interposed between two joint surfaces that are covered by hyaline cartilage. The pubic symphysis, for example, joins the two pubic bones. Normally, this joint is relatively immovable, except during late pregnancy and birth when hormonal changes allow greater tissue extensibility and joint movement. Other examples of symphyses include the sternomanubrial joint (joining the manubrium with the sternal body) and intervertebral joints (between adjacent vertebrae).

Synovial Joints

Most movements of our extremities occur at major joints such as the hip, knee, ankle, shoulder, elbow, and wrist. All of these joints, and many others, are synovial joints.

Synovial joints share common structural elements, including a **synovial joint cavity** encapsulated by a **fibrous joint capsule**. The joint cavity is filled with **synovial fluid** produced by the **synovial membrane**, a thin membrane that lines the inner surface of the joint capsule. A smooth, shiny layer of **articular cartilage** (composed of the hyaline type of cartilage) covers the joint surfaces of the articulating bones. This structure allows the joints to move freely. Synovial joints therefore are functionally classified as diarthroses.

The fibrous joint capsule consists of dense connective tissue that blends with the periosteum of the articulating bones (figure 3.2). The capsule forms the outer boundary of the synovial joint and provides structural support. A thin synovial membrane covers the inner surface of the capsule. Although not providing any structural support, the synovial membrane is nonetheless important for overall joint function because it secretes a thick synovial fluid that lubricates the joint.

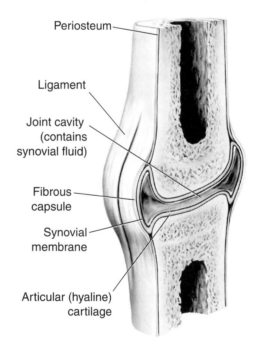

Periosteum

Ligament

Joint cavity (contains synovial fluid)

Fibrous capsule

Synovial membrane

Articular (hyaline) cartilage

Figure 3.2   Structure of a synovial joint.

As a lubricant, the fluid reduces friction between the articular surfaces and assists with the absorption of compressive loads applied to the joint. The synovial fluid also provides nutrients for the articular cartilage. The hyaline cartilage on the joint surfaces is avascular (i.e., lacks an inherent blood supply) and therefore must rely on the synovial fluid to provide nutrients through diffusion. Synovial joints contain a relatively small amount of this important fluid. Large joints such as the knee, for example, contain a scant 3 ml or less of synovial fluid.

When irritated, as in joint trauma, the synovial membrane produces excess synovial fluid. This excess fluid causes the joint to swell. A severely twisted knee, for example, may swell in response to injury and be diagnosed as so-called water on the knee. In fact, the swelling results from overproduction of synovial fluid by an irritated synovial membrane.

Joint Cracking

Some joints, particularly those whose joint surfaces are congruent (i.e., fit well together), can produce a "cracking" or "popping" sound when pulled apart or quickly forced toward their end range of motion. Two questions often arise. First, what causes the popping sound? Second, does cracking knuckles cause injury to the joint? The popping sound is caused by a change in hydrostatic pressure in the fluid inside synovial joints. When the bones are first pulled, a negative pressure develops in the synovial fluid, and a bubble of carbon dioxide gas forms. As the bones separate, the pressure within the joint causes the bubble to collapse and "pop." The collapsed bubble dissolves into smaller bubbles, which eventually disappear as the joint's fluid state returns to normal. As for the second question, the hydrostatic pressure change and popping sound will not, in and of themselves, cause injury. Rapid and repeated stretching of articular structures, however, may lead to injury if the tissues are deformed to the extremes of joint motion.

Synovial joints may also contain other structural components that enhance their function. These accessory structures include ligaments, tendons, bursae, tendon sheaths, fibrocartilage pads (menisci), and fat pads.

The tensile strength of ligaments makes them ideal structures to reinforce synovial joints and enhance their resistance to dislocation. Three ligament types are found in and around synovial joints. The most common type, the capsular ligament, blends with the joint capsule and appears as a capsular thickening. Extracapsular ligaments also are common and lie completely outside the joint capsule. Much less common are intracapsular ligaments that lie entirely within the capsule and attach directly to the bone.

Tendons, although technically not a part of the joint itself, can lend structural support through their action across or around joints. Muscle forces transmitted through the tendon to the bone can reinforce the joint and increase its stability. The contribution of muscle action to joint stability is sometimes termed **active support**, in contrast to the **passive support** provided by noncontractile (i.e., non-force-producing) tissues such as ligaments and the joint capsule.

Bursae are sacs, or pockets, filled with synovial fluid that serve as spacers between a tendon or ligament and the underlying bone. They reduce friction and pressure and provide shock absorption. Tendon sheaths are elongated, tube-shaped bursae that surround tendons and facilitate tendon movement by providing a lubricated surface. Joints are further supported and cushioned by fat pads that act as fillers around the joint.

Articular discs, or **menisci**, are fibrocartilage pads interposed between bones. These pads provide joint cushioning and act as wedges to improve the bony fit. They typically are ring- or C-shaped and attach to one of the articular surfaces. The menisci of the knee (figure 3.3), for example, attach to the joint surfaces of the proximal tibia and are necessary for effective knee joint function.

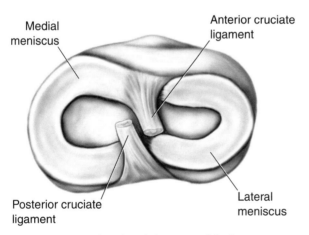

Medial meniscus

Anterior cruciate ligament

Posterior cruciate ligament

Lateral meniscus

Figure 3.3 Lateral and medial menisci of the knee.

Classification of Synovial Joints

Bones come in a wide variety of shapes and sizes. The bones of each joint are uniquely configured to permit certain movements and restrict others. Synovial joints are structurally classified according to their respective bony fit and movement potential. Each of the six classification types has mechanical correlates, as illustrated in figure 3.4.

Arthritis

Arthritis refers to inflammation of a joint. The term *arthritis* encompasses many conditions that have either primary or secondary inflammatory involvement. Among the major types are those resulting from chronic and excessive mechanical loading (e.g., osteoarthritis), systemic disease (e.g., rheumatoid arthritis), or biochemical imbalances (e.g., gouty arthritis).

Osteoarthritis (OA), also known as degenerative joint disease, is the most common form of arthritis. It is characterized by deterioration of the hyaline articular cartilage and by bone formation on joint surfaces. Osteoarthritis results from mechanical trauma and accompanying chemical process alterations. Given its mechanical etiology, OA most often affects the load-bearing joints of the lower extremities (e.g., hip and knee).

Rheumatoid arthritis (RA) is an autoimmune condition in which the body's immune system malfunctions and attacks its own joint tissues. RA often affects the joints of the hand and fingers and is characterized by considerable joint swelling, pain, and limited range of motion. In severe cases, the bones can become fused together. This eliminates joint motion altogether.

Gouty arthritis (or gout) is caused by excess production of uric acid in the form of uric acid crystals. These crystals are carried by the circulatory system and become embedded in synovial joint structures. The crystals irritate the joint and initiate an inflammatory response.

All types of arthritis can cause debilitating pain and functional loss at joints of the musculoskeletal system.

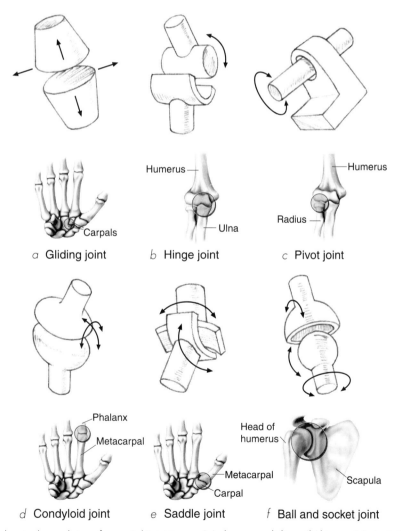

Figure 3.4 Mechanical correlates of synovial joint types: *(a)* planar or gliding, *(b)* hinge, *(c)* pivot, *(d)* condyloid, *(e)* saddle, *(f)* ball and socket.

Reprinted, by permission, from P. McGinnis, 2004, *Biomechanics of sport and exercise,* 2nd edition (Champaign, IL: Human Kientics), 245.

- *Gliding (planar) joints* have opposing flat or slightly curved surfaces that permit limited sliding between bones (figure 3.4a). These joints have no axis of rotation, and their movements are described as planar. Gliding joints include those formed by the articular processes of adjacent vertebrae, intertarsal joints, and intercarpal joints.

- *Hinge joints* exhibit angular motion about a single fixed axis of rotation, much like that of a door hinge (figure 3.4b). Movements are restricted to a single plane (uniplanar) and a single axis (uniaxial) by the joint's bony configuration. These joints (e.g., ankle, elbow, interphalangeal) are relatively stable, largely because of their tight bony fit.

- *Pivot joints* are uniaxial (like hinge joints) but are distinguished by an axis that runs longitudinally along a bone (figure 3.4c). The pivot joint between the atlas and axis (in the neck) permits rotation of the head from side to side. Similarly, the proximal radioulnar joint pivots to allow the radius to rotate (or "roll") over the ulna in pronation of the forearm.

- *Condyloid (ellipsoid) joints* are relatively unstable joints formed by articulation of a shallow convex surface of one bone with the concave surface of another (figure 3.4d). The metacarpophalangeal and wrist joints provide good examples of condyloid joints and their ability to move in two planes (biplanar). The articulation between the distal femur and the proximal tibia at the knee forms a double condyloid (or bicondyloid) joint. Given the instability of this configuration, the knee is reinforced by numerous ligaments and other supporting structures. (Note that the knee sometimes is classified as a hinge joint; this is technically incorrect. Although the primary movement at the knee involves flexion or extension, the knee also can rotate when in a flexed position and hence is biplanar. This violates the uniplanar requirement of a hinge joint.)

- The name *saddle joint* is given to articulations where one bone sits on another as a saddle sits on a horse (figure 3.4e). The concave surface of one bone ("the saddle") straddles the convex surface of another bone ("the horse's back"). Generally considered biplanar and biaxial, saddle joints are relatively stable because of the interlocking of the two bones. The carpometacarpal joint of the thumb is an example of a saddle joint. When you "twiddle your thumbs," the biplanar nature of this saddle joint becomes obvious.

- *Ball-and-socket joints* are formed when the rounded end of one bone is housed in a depression (socket) of another bone (figure 3.4f). Ball-and-socket joints, such as the hip and shoulder, are triplanar and triaxial and demonstrate tremendous range of motion.

Joint Motion and Movement Description

A full appreciation of human movement requires an understanding of anatomical details and certain movement concepts. Some of these concepts have been outlined in earlier sections, and we build on them here with more depth.

There are a few key points to bear in mind when considering joints and motion. Specific movement-related descriptors, such as *flexion* and *extension,* can be used in one of two ways. They may describe *movements* (e.g., "Maria *flexed* her elbow") or *positions* (e.g., "Trevor maintained a *flexed* elbow while carrying the box"). The context usually makes clear whether a specific term is being used to describe a movement or a position. Also keep in mind that all movements are described from the mover's perspective. Consider the case of a woman, for example, who curls (flexes) her right elbow in lifting a dumbbell. Viewed face to face, the movement occurs on the observer's left side. Described from the mover's perspective, however, the movement is flexion of the right elbow.

Joint Stability and Mobility

Some joints, such as the hip, are very stable (i.e., they strongly resist dislocation), while others such as the glenohumeral joint are unstable (i.e., they are dislocated fairly readily). From a movement perspective, some joints move freely and have extensive mobility, while others are

Double-Jointed Joints

The term *double jointed* is sometimes used to describe joints that are hypermobile, allowing extreme range of motion. Contrary to what the name might suggest, double-jointed articulations do not contain a second joint but rather have shallow bony articulations, loose joint capsules, and lax ligaments. These structural characteristics allow greater than normal range of motion and possible **subluxation** (i.e., partial dislocation) of the joint.

immovable, or immobile. In general, very stable joints are relatively immobile; conversely, relatively unstable joints typically are very mobile. This concept describes a stability–mobility continuum: joints built for stability are less mobile, and joints designed for mobility are less stable. The concept of a stability–mobility continuum holds for most joints. Where a given joint falls on this continuum depends on the degree of bony fit, joint capsule tightness, the amount of ligamentous support, and whether the joint contains a stabilizing fibrocartilage structure such as a meniscus. The presence or absence of one or more of these characteristics, however, does not ensure a joint's relative stability or mobility. A strong complement of ligaments and a tight joint capsule, for example, might compensate for poor bony fit and result in a relatively stable joint.

As with most rules, however, there are exceptions. The hip joint, for example, is both mobile and stable. The hip's mobility is afforded by its ball-and-socket construction, with the head of the femur fitting into the socket (acetabulum) of the pelvis; its stability is provided by the considerable reinforcement of periarticular tissues, especially the large muscle mass surrounding the joint. The depth of the femoral head within the acetabulum contrasts with the large humeral head and shallow glenoid of the unstable glenohumeral (shoulder) joint.

Movement Planes and Joint Motion

If a person moves a body segment away from anatomical position (figure 1.6, page 18), subsequent movements of that segment may occur in a plane different from the one in anatomical position. For example, when in anatomical position, external (or lateral) rotation of the hip joint occurs in the transverse plane. If the hip is flexed (from anatomical position) 90° before it is externally rotated, the rotation now happens in the frontal plane (figure 3.5a).

As another example, consider the elbow joint. From anatomical position, elbow flexion and extension occur in the sagittal plane. If the glenohumeral joint is first abducted 90°, subsequent elbow flexion and extension now happen in the transverse plane (figure 3.5b).

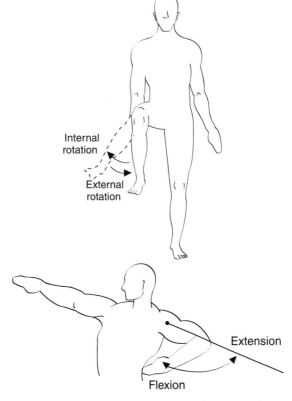

Figure 3.5 Movements occurring out of anatomical position: (a) after hip flexion, internal and external rotation of the thigh now occurs in the frontal plane, as compared with its movement in the transverse plane when in anatomical position; (b) after abduction of the arm, elbow flexion and extension now occur in the transverse plane, as compared with its movement in the sagittal plane when in anatomical position.

Contortionists

Contortionism has been defined as "the art of manipulating the body parts in feats of extreme suppleness and skill" (Alter, 1996, p. 110). As shown in figure 3.6, contortionists have joint hypermobility, and some can even voluntarily dislocate joints to achieve curious postures. As to the question of whether contortionists are "born" or "made," the answer would appear to be both. Some performers seem genetically predisposed to hypermobility and can achieve contorted positions with little training, while others must spend several hours a day training to maintain their remarkable flexibility and joint laxity (Alter, 1996).

Figure 3.6   Contortionists.

Movements confined to a single plane often appear restricted and robotic. Full expression of the wide variety and elegance of human movement depends on our ability to move unconstrained in three dimensions. The amount of movement allowed at a particular joint is dictated by its structure and described by the number of planes in which a segment can move or the number of primary rotational axes a joint possesses.

A uniplanar joint (e.g., elbow) has a single axis of rotation. Biplanar and biaxial articulations such as the metacarpophalangeal joints move in two planes. Triplanar joints are free to move in three planes. The hip and glenohumeral joints, for example, move in three planes: flexion and extension in the sagittal plane, abduction and adduction in the frontal plane, and internal and external rotation in the transverse plane.

Types of Joint Movement

As just discussed, the bony structure and periarticular tissues of synovial joints dictate movement potential at each joint. In this section we consider four types of joint movement. Remember that joint movements are described according to the type of movement (e.g., gliding, rotation) and the plane of movement (sagittal, frontal, or transverse) relative to the standard reference of anatomical position.

Gliding and Angular Movements

The simple sliding, or gliding, of two surfaces on one another is considered uniplanar and typically does not involve any rotation. Gliding movements are seen at the intercarpal and intertarsal joints, where the amount of gliding is very limited owing to the tightness of the joint capsule and supporting ligaments.

Angular motion occurs when a body segment moves through an angle about an imaginary line called the axis of rotation. The axis of rotation is usually located in either the proximal or distal end of the segment. When the forearm is rotated about the elbow during a biceps curl, the axis of rotation is a line through the elbow that is perpendicular to the plane of motion (figure 3.7a).

Flexion is angular movement occurring in the sagittal plane (relative to anatomical position) in which the angle between articulating segments decreases. **Extension** is the opposite

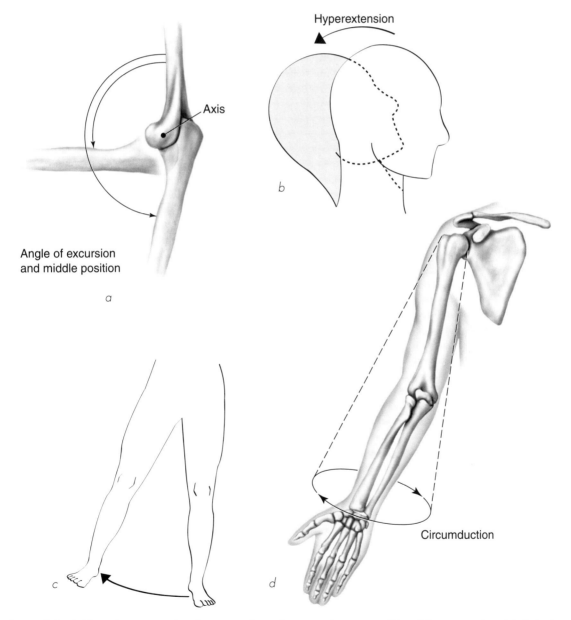

Figure 3.7 *(a)* Axis of rotation at the elbow joint during flexion and extension, *(b)* neck hyperextension, *(c)* abduction of the leg from anatomical position, *(d)* circumduction.

movement when the relative joint angle increases. For example, the neck flexes when the head nods forward, with the chin moving toward the chest. Neck extension returns the head to its upright position. When a joint angle increases beyond anatomical position, the movement is termed **hyperextension** (figure 3.7b), such as when the head is tilted backward from anatomical position. Only certain joints can hyperextend without causing injury. The glenohumeral and hip joints easily swing back posteriorly into hyperextension. The elbow and knee, however, risk serious injury if hyperextended.

In the frontal plane, angular movement that takes a segment away from the body's midline is termed **abduction**. This happens, for example, when the leg is lifted out to the side from anatomical position (figure 3.7c). Moving the segment back toward the midline is called **adduction**. Abduction and adduction also describe movement of the fingers relative to the midline of the hand. Spreading the fingers apart is abduction; bringing them back together is adduction. The terms *abduction* and *adduction* are used only to describe movements of the appendicular skeleton.

A potentially confusing situation arises when the arm is elevated to the side (abducted) beyond 90°. As the arm moves past 90° to an overhead position, it moves back toward the body's midline, and the movement might be called adduction. While technically correct, we do not describe the motion that way. Frontal plane movement from anatomical position to overhead is called abduction throughout the full range of motion. Similarly, the return movement from an overhead position back to anatomical position is termed adduction.

Circumduction is a special form of angular motion in which the distal end of a limb or segment moves in a circular pattern about a relatively fixed proximal end. In three-dimensional space, circumduction traces out a cone-shaped pattern (figure 3.7d). Examples of circumduction include rotating the arm in full circles about the shoulder and moving a finger in a circular pattern about its proximal metacarpophalangeal joint.

Rotational Movements

Rotational movement, or rotation, is distinguished from angular movement by the fact that the axis of rotation is oriented along the long axis of a bone or segment instead of passing through one end of the segment (figure 3.8). From anatomical position, rotation occurs in the transverse plane. (Note that the term *rotation* is technically defined as above, but also is used to describe angular movement of a segment about its joint axis [e.g., hip rotation].)

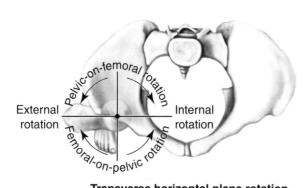

Transverse horizontal plane rotation

Figure 3.8 Joint rotation about a longitudinal axis and internal and external rotation of the hip.

Adapted from *Kinesiology of the Musculoskeletal System*, D.A. Neumann, pg. 402, Copyright 2002, with permission from Elsevier.

Rotations of the axial skeleton (e.g., when the spine twists from side to side) are simply described as rotation left or rotation right. (Remember, the directions *left* and *right* are described relative to the person moving, not from the perspective of someone observing the movement.)

Rotational movements in appendicular segments are referenced by their direction relative to the body's midline. The anterior surface of the segment (in anatomical position) is used as the reference. If movement rotates the segment's anterior surface inward toward the midline, the movement is termed **internal** (or **medial**) **rotation**. Rotational movement away from the midline of the body is called **external** (or **lateral**) **rotation** (see figure 3.5a).

Special Movements

The general movement forms just discussed describe most movements. Some movements, however, have special names unique to a particular joint or body region.

• At the ankle, movement of the foot away from the lower leg is called **plantar flexion**, as when performing a calf raise exercise or depressing an automobile's gas pedal. The reverse movement of bringing the foot upward toward the lower leg is termed **dorsiflexion** (figure 3.9).

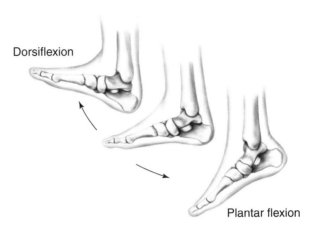

Figure 3.9 Plantar flexion and dorsiflexion of the ankle.

• **Inversion** is a movement of the intertarsal joints (particularly the talocalcaneal, or subtalar, joint) that results in the sole of the foot being moved inward toward the midline of the body. For **eversion**, the sole of the foot is moved away from the body's midline (figure 3.10). From anatomical position, eversion and inversion happen primarily in the frontal plane.

• Simultaneous rotation of the proximal and distal radioulnar joints (in the forearm) allows the radius (lateral forearm bone) to rotate over the relatively stationary ulna (medial forearm bone). From anatomical position, rotation of the radius over the ulna is termed **pronation**. The return from a pronated position back to anatomical position is called **supination**. In anatomical position, the radioulnar joints place the forearm in what is termed a supinated position. When shaking hands, the forearm is in midposition, halfway between

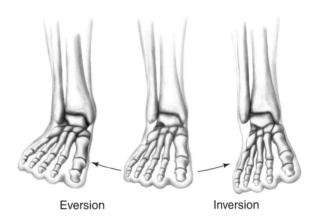

Figure 3.10 Eversion and inversion of the foot.

the pronated and supinated positions. (Important note: Many authors, though certainly not all, also use the terms *supination* and *pronation* to describe combined movements of the foot and ankle. Supination of the foot and ankle commonly, though not universally, describes the combination of inversion, plantar flexion, and internal [medial] rotation, or adduction, of the foot. Pronation of the foot and ankle describes the opposite combination of eversion, dorsiflexion, and external [lateral] rotation, or abduction, of the foot.)

• In an earlier section, abduction and adduction were used to describe movements in the frontal plane away from the midline and toward the midline, respectively. Abduction of the wrist and hand, such that the thumb moves closer to the radius, is termed **radial deviation**. Adduction of the wrist and hand, such that the little (fifth) finger moves closer to the ulna, is called **ulnar deviation**.

• **Opposition** refers to the thumb's ability to work with the other four fingers to perform grasping movements. This ability is of paramount importance in allowing us to manipulate objects (e.g., picking up a pencil, turning a key).

• **Elevation** refers to a structure moving in a superior, or upward, direction. **Depression** describes an inferior, or downward, movement. These terms are typically used to describe movements of the jaw (mandible) and scapulae. Elevation is seen when closing the mouth or shrugging the shoulders. Depression results in opening the mouth and dropping the shoulders. These movements usually occur in the frontal plane.

• **Protraction** describes movement anteriorly, or toward the front of the body. **Retraction** refers to posteriorly directed movement toward the back. These movements occur in the transverse plane. For example, the clavicles and scapulae protract when one slumps forward with rounded shoulders and crossed arms. These bones retract when the shoulders are pulled back into an upright and correct posture.

- **Lateral flexion** (right and left) refers to sideways bending of the vertebral column in the frontal plane.
- Upward and downward rotation refers to rotation of the scapulae.

Joint Structure and Movement

Gross movements of the human musculoskeletal system are made primarily by the major synovial joints of the extremities, along with contributions from joints of the head, neck, and spine. This section explores each joint's structure and movements. A summary of joint structure and movements is given in tables 3.1, 3.2, and 3.3. Typical range of motion values are shown in table 3.4.

Table 3.1   **Summary of Joint Structure and Movement (Pelvis and Lower Extremity)**

| Joint | Structural classification | [All movements begin from anatomical position] | | |
		Movement	Plane	Axial/Planarity
Sacroiliac	Synovial (plane)	Gliding		Nonaxial/ Nonplanar
Pubic symphysis	Symphysis	Distraction/Separation during birth		
Pelvic girdle (movement of pelvis relative to femur)	Synovial (ball and socket)	Anterior tilt Posterior tilt	Sagittal	Triaxial/Triplanar
		Lateral tilt right Lateral tilt left	Frontal	
		Rotation right Rotation left	Transverse	
Hip (movement of femur relative to pelvis)	Synovial (ball and socket)	Flexion Extension Hyperextension	Sagittal	Triaxial/Triplanar
		Abduction Adduction	Frontal	
		Internal (medial) rotation External (lateral) rotation	Transverse	
	[starting with hip flexed 90°]	Horizontal abduction (horizontal extension) Horizontal adduction (horizontal flexion)	Transverse	
Patellofemoral	Synovial (plane)	Gliding		Nonaxial/ Nonplanar
Tibiofemoral (knee)	Synovial (bicondyloid)	Flexion Extension	Sagittal	Biaxial/Biplanar
		Internal (medial) rotation External (lateral) rotation [with knee flexed]		

Joint	Structural classification	[All movements begin from anatomical position] Movement	Plane	Axial/Planarity
Ankle	Synovial (hinge)	Dorsiflexion Plantar flexion	Sagittal	Uniaxial/ Uniplanar
Subtalar	Synovial (plane)	Inversion Eversion	Frontal	Uniaxial/ Uniplanar
Intertarsal	Synovial (plane)	Gliding		Uniaxial/ Uniplanar
Tarsometatarsal	Synovial (plane)	Gliding		Nonaxial/ Nonplanar
Metatarsophalangeal	Synovial (condyloid)	Flexion Extension Hyperextension	Sagittal	Biaxial/Biplanar
		Abduction Adduction	Transverse	
Interphalangeal	Synovial (hinge)	Flexion Extension	Sagittal	Uniaxial/ Uniplanar

Table 3.2 Summary of Joint Structure and Movement (Upper Extremity)

Joint	Structural classification	[All movements begin from anatomical position] Movement	Plane	Axial/Planarity
Sternoclavicular (shoulder girdle)	Synovial (ball and socket)	Anterior rotation Posterior rotation	Sagittal	Triaxial/Triplanar
		Upward rotation Downward rotation	Frontal	
		Abduction Adduction	Transverse	
Acromioclavicular	Synovial (plane)	Gliding		Nonaxial/ Nonplanar
Glenohumeral (shoulder)	Synovial (ball and socket)	Flexion Extension Hyperextension	Sagittal	Triaxial/Triplanar
		Abduction Adduction	Frontal	
		Internal (medial) rotation External (lateral) rotation		
	[starting with shoulder flexed 90°]	Horizontal abduction (horizontal extension) Horizontal adduction (horizontal flexion)	Transverse Transverse	
Elbow	Synovial (hinge)	Flexion Extension	Sagittal	Uniaxial/ Uniplanar
Radioulnar	Proximal: synovial (pivot) Middle: syndesmosis Distal: synovial (pivot)	Pronation Supination	Transverse	Uniaxial/ Uniplanar

(continued)

Table 3.2 Summary of Joint Structure and Movement (Upper Extremity) *(continued)*

| Joint | Structural classification | [All movements begin from anatomical position] | | |
		Movement	Plane	Axial/Planarity
Radiocarpal (wrist)	Synovial (condyloid)	Flexion Extension Hyperextension	Sagittal	Biaxial/Biplanar
		Radial deviation (abduction) Ulnar deviation (adduction)	Frontal	
Intercarpal	Synovial (plane)	Gliding		Nonaxial/ Nonplanar
Carpometacarpal	Synovial (plane)	Gliding		Nonaxial/ Nonplanar
Metacarpophalangeal	(1) Thumb: synovial (saddle) (2-5): Synovial (condyloid)	Flexion Extension Hyperextension	(1) Frontal (2-5) Sagittal	Biaxial/Biplanar
		Abduction Adduction	(1) Sagittal (2-5) Frontal	
Interphalangeal	Synovial (hinge)	Flexion Extension	Sagittal	Uniaxial/ Uniplanar

Table 3.3 Summary of Joint Structure and Movement (Head, Neck and Trunk)

| Joint | Structural classification | [All movements begin from anatomical position] | | |
		Movement	Plane	Axial/Planarity
Intercranial	Suture	None		
Temporomandibular	Synovial (condyloid)	Elevation Depression	Sagittal	Biaxial/Biplanar
		Protraction Retraction	Transverse	
Atlantooccipital	Synovial (hinge)	Flexion Extension	Sagittal	Uniaxial/ Uniplanar
Vertebral Column: Atlantoaxial	Synovial (pivot)	Rotation right Rotation left	Transverse	Uniaxial/ Uniplanar
C2-L5: (Vertebral bodies: symphysis) (Articular processes: synovial [plane])		Flexion Extension Hyperextension	Sagittal	Triaxial/Triplanar
		Lateral flexion right Lateral flexion left	Frontal	
		Rotation right Rotation left	Transverse	
Costovertebral	Synovial (plane)	Gliding		Nonaxial/ Nonplanar
Sternomanubrial	Symphysis	Sternal angle increase Sternal angle decrease		Nonaxial/ Nonplanar

Table 3.4   Average Ranges of Joint Motion*

Joint	Joint motion	ROM (degrees)
Hip	Flexion	90-125
	Hyperextension	10-30
	Abduction	40-45
	Adduction	10-30
	Internal rotation	35-45
	External rotation	45-50
Knee	Flexion	120-150
	Rotation (when flexed)	40-50
Ankle	Plantar flexion	20-45
	Dorsiflexion	15-30
Shoulder	Flexion	130-180
	Hyperextension	30-80
	Abduction	170-180
	Adduction	50
	Internal rotation**	60-90
	External rotation**	70-90
	Horizontal flexion**	135
	Horizontal extension**	45
Elbow	Flexion	140-160
Radioulnar	Forearm pronation (from midposition)	80-90
	Forearm supination (from midposition)	80-90
Cervical spine	Flexion	40-60
	Hyperextension	40-75
	Lateral flexion	40-45
	Rotation	50-80
Thoracolumbar spine	Flexion	45-75
	Hyperextension	20-35
	Lateral flexion	25-35
	Rotation	30-45

*Range of motion (ROM) for movements made from anatomical position (unless otherwise noted). Averages reported in the literature vary, sometimes considerably, depending on method of measurement and population measured. Values above are representative of the ranges of reported maximum ROM.

**Movement from abducted position.

Movements of the Hip and Pelvis

The hip joint is formed by articulation of the head of the femur with the acetabulum of the pelvis (figure 3.11). Also known as the coxofemoral (or coxal) joint, the hip is a synovial ball-and-socket joint capable of movement in all three primary planes (i.e., triplanar). As described earlier in the chapter, the hip joint is relatively stable and resists dislocation, largely due to the good fit of the femoral head in the acetabulum and the large muscle mass surrounding the joint. The joint's stability

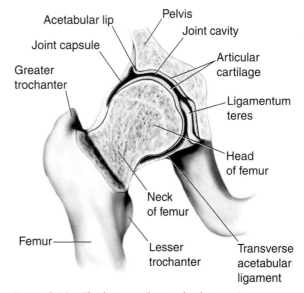

Figure 3.11   The hip joint (longitudinal section).

is enhanced by its **labrum**, a U-shaped ring of fibrocartilage around the rim of the acetabulum. The labrum deepens the socket (acetabulum) and thereby improves the joint's bony fit.

Relative movement between the pelvis and femur can be viewed in two ways: (1) with the pelvis fixed, the femur is free to move through all three planes; and (2) with the femur fixed, the pelvis can perform triplanar movement.

In the first case, with a fixed pelvis, the femur can move in flexion and extension (sagittal plane, figure 3.12a), abduction and adduction (frontal plane, figure 3.12b), and internal (medial) and external (lateral) rotation (transverse plane, figure 3.12c). A circular, or cone-shaped, movement pattern known as circumduction combines the movements of flexion, extension, abduction, and adduction.

When the femur is fixed, as in standing, and the pelvis is allowed to move, possible movements include anterior and posterior tilt in the sagittal plane (figure 3.13a), lateral tilt left and right in the frontal plane (figure 3.13b), and rotation left and right in the transverse plane (figure 3.13c).

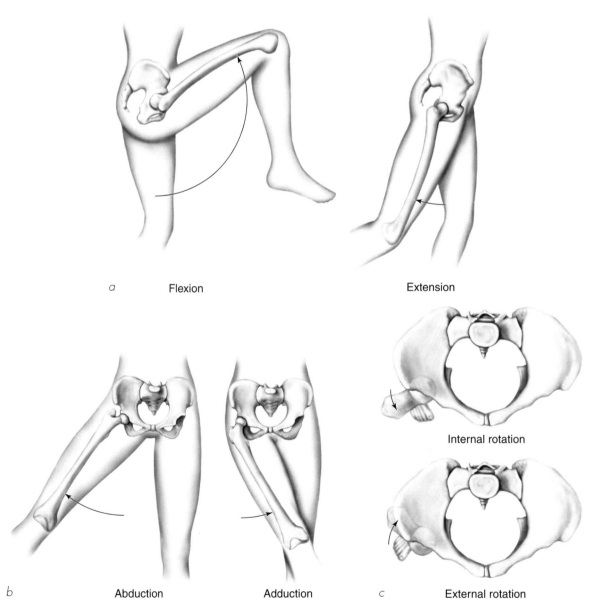

a   Flexion   Extension

                                                Internal rotation

b   Abduction   Adduction   c   External rotation

Figure 3.12 Hip movement about a fixed pelvis: (a) flexion and extension, (b) abduction and adduction, (c) internal and external rotation.

Adapted from *Kinesiology of the Musculoskeletal System*, D.A. Neumann, pg. 403, Copyright 2002, with permission from Elsevier.

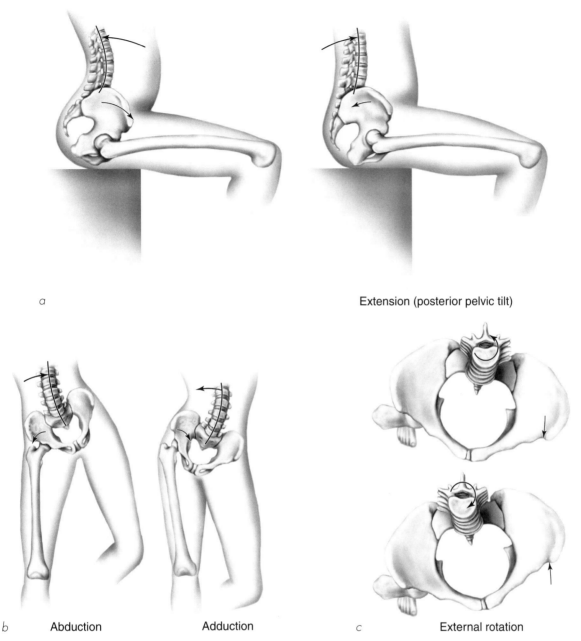

a   Extension (posterior pelvic tilt)

b   Abduction   Adduction   *c*   External rotation

Figure 3.13 Pelvic movement about a fixed femur: *(a)* anterior and posterior tilt, *(b)* lateral tilt right and left, *(c)* rotation right and left.

Adapted from *Kinesiology of the Musculoskeletal System,* D.A. Neumann, pg. 405, Copyright 2002, with permission from Elsevier.

Movements of the Knee

The knee (tibiofemoral) joint, the largest joint in the body, is formed by articulation of the medial and lateral condyles at the distal end of the femur with matching condylar surfaces at the proximal end of the tibia (figure 3.14a). The knee joint is the most complex of the major synovial joints. Although it functions as a modified hinge joint, the knee joint is more commonly referred to as a bicondyloid (or double condyloid) joint. The articulation of the femoral condyles with the tibial surfaces is enhanced by the presence of two menisci (medial and lateral). The medial and lateral menisci (figure 3.14b) are semicircular rings of fibrocartilage that deepen the articular surface, improve the bony fit, and stabilize the knee joint.

The anterior surface of the distal femur also articulates with the patella (kneecap) to form the patellofemoral joint. The patella improves the leverage of the muscles (quadriceps group) responsible for extending the knee by moving the muscles' line of action away from the knee joint's axis of rotation.

The knee acts primarily as a modified hinge joint, with flexion and extension happening in the sagittal plane (figure 3.15). As the knee flexes, however, some rotation is possible between

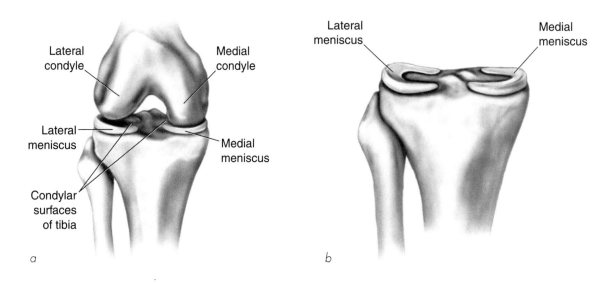

Figure 3.14 Knee: bones and joints.

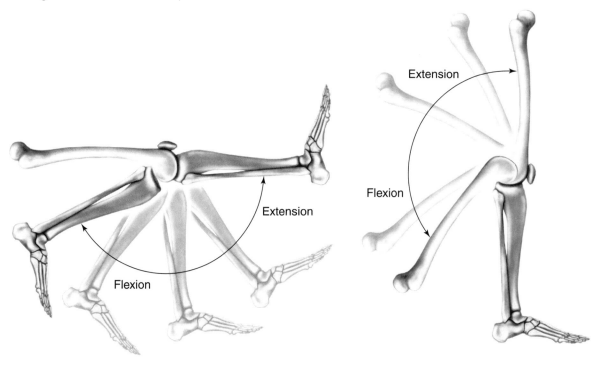

Tibial-on-femoral perspective   **Femoral-on-tibial perspective**

Figure 3.15   Knee joint movements in the sagittal plane: tibial motion about a fixed femur and femoral movement about a fixed tibia.

Adapted from *Kinesiology of the Musculoskeletal System,* D.A. Neumann, pg. 444, Copyright 2002, with permission from Elsevier.

the lower leg (tibia) and the thigh (femur). During the final few degrees of extension (as the knee approaches full extension), the tibia and femur rotate relative to one another to prevent rotation of the lower leg. This tibiofemoral rotation, termed the **screw-home mechanism**, stabilizes the knee at full extension for walking and running.

At the patellofemoral joint, the patella slides into the intercondylar groove of the femur as the knee moves from full extension into flexion. This movement is termed **patellar tracking**. In a well-functioning knee, the patella tracks in the middle of the groove. In some cases, however, conditions such as injury, muscle weakness, or paralysis may cause the patella to track to one side of the groove, usually to the lateral side. This maltracking can be painful and limit performance of tasks involving the knee (e.g., running and jumping).

Movements of the Ankle and Foot

The numerous bones, ligaments, and articulations in the ankle and foot region make this one of the body's most structurally complex areas. The ankle (talocrural) joint is a synovial joint formed by articulation of the distal ends of the tibia and fibula with the superior surface of the talus. The talus fits into a deep socket, or mortise, formed by the tibia and fibula. The ankle functions as a hinge joint, with the talus rotating between the tibial and fibular malleoli. In a dorsiflexed position, the talus fits snugly within the mortise, and the ankle joint is very stable. As the ankle plantar flexes, a narrower portion of the talus rotates into the mortise. This results in a looser bony fit, compromising joint stability; the ankle therefore is relatively unstable in the plantar flexed position.

As described in chapter 2, each foot contains 26 bones, including 7 tarsals, 5 metatarsals, and 14 phalanges (figure 2.6, page 31). Each bone articulates with one or more adjacent bones. Joints between neighboring tarsal bones are termed intertarsal joints, in general, with specific joints identified by the involved bones (e.g., cuboideonavicular joint).

One of the most important intertarsal joints is the articulation of the superior surface of the calcaneus with the inferior surface of the talus. This talocalcaneal, or subtalar, joint plays an essential role in proper functioning of the foot and ankle complex during load-bearing activities such as walking and running. The subtalar joint axis runs obliquely, as shown in figure 3.16.

The distal tarsal bones (cuboid and cuneiforms) join with the proximal ends of the metatarsals to form five tarsometatarsal joints. Similarly, the distal ends of the metatarsals articulate with heads of the proximal phalanges to create five metatarsophalangeal (MP) joints. Finally, interphalangeal (IP) joints are hinge joints formed between adjacent phalanges in each of the toes.

The primary ankle movements are dorsiflexion and plantar flexion. Relative to anatomical position, both movements occur in the sagittal plane. Dorsiflexion describes movement of the dorsal (top) surface of the foot toward the lower leg (see figure 3.9). The opposite movement, plantar flexion, involves movement of the plantar (bottom) surface of the foot away from the lower leg. Standing on one's tiptoes and pushing down on the accelerator pedal in a car involve ankle plantar flexion.

The subtalar joint is primarily responsible for inversion and eversion. Inversion involves a tilting of the foot in the frontal plane so that the bottom of the foot faces medially (see figure

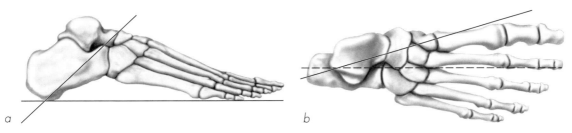

a   b

Figure 3.16   Subtalar joint axis, inclined (a) superiorly and (b) medially.

3.10). In eversion, the foot tilts so that the sole of the foot faces laterally. Inversion and eversion both happen primarily at the subtalar joint.

As noted earlier, the terms *supination* and *pronation* are commonly (though not universally) used to describe combined movements of the ankle and foot. Supination of the foot and ankle describes the combination of ankle plantar flexion, subtalar inversion, and internal (medial) rotation of the foot. Pronation of the ankle and foot describes the opposite combination of ankle dorsiflexion, subtalar eversion, and external (lateral) rotation of the foot.

In addition to the subtalar joint, other intertarsal articulations contribute to a lesser degree to inversion and eversion. These joints include the calcaneocuboid joint and the talonavicular portion of the talocalcaneonavicular joint. In general, the tight fit between the intertarsal bones largely restricts them to slight gliding relative to adjacent bones.

The tarsometatarsal articulations are synovial joints limited to slight gliding movements. The metatarsophalangeal joints, as condyloid synovial joints, allow flexion, extension, and hyperextension in the sagittal plane, along with abduction and adduction in the transverse plane. The interphalangeal joints of the toes act as hinge joints allowing flexion and extension in the sagittal plane.

Movements of the Shoulder

 The shoulder complex includes joints involving the scapula, sternum, clavicle, and humerus. The clavicle attaches medially to the sternal manubrium at the sternoclavicular joint, a synovial joint with a fibrocartilage disc separating the bony surfaces.

At its lateral end, the clavicle articulates with the acromion process of the scapula at the acromioclavicular (AC) joint. The AC joint is a gliding synovial joint with articular surfaces separated by an articular disc.

The humerus of the upper arm articulates with the scapula at the glenohumeral (GH) joint (also shoulder joint). The GH joint is the body's most mobile joint, where the humeral head fits loosely into the shallow glenoid fossa of the scapula. The shoulder's loose joint capsule provides little stability to the joint and explains why the glenohumeral articulation is a frequent site of joint dislocation. A fibrocartilage glenoid labrum attaches to the rim of the glenoid fossa and improves the joint's bony fit.

The glenohumeral joint is a triplanar articulation whose ball-and-socket structure allows it the greatest mobility of any joint in the body. In the sagittal plane, the arm flexes and extends (3.17a). Abduction and adduction occur in the frontal plane (3.17b), with internal (medial) rotation and external (lateral) rotation happening in the transverse plane (figure 3.17c). When the upper arm is medially rotated, the range of abduction is limited to about 60° because of pinching (impingement) of the greater tubercle of the humerus on the acromion process of the scapula. Abduction beyond this point requires external rotation of the upper arm, which frees the tubercle from its acromial restriction and permits further movement.

In addition to these movements in the primary planes, the arm at the GH joint can move in horizontal abduction (also horizontal extension) and horizontal adduction (also horizontal flexion) (figure 3.17d). Also, in the same manner as the leg acting at the hip, the arm can move in a circular, or conical, pattern known as circumduction.

The scapula is primarily anchored by muscles and thus can be described as being in muscular suspension. Its only bony articulations are with the humerus and clavicle. The many muscles attaching to the scapula dictate its movement. These movements include elevation and depression, retraction and protraction, and downward and upward rotation (figure 3.18).

Glenohumeral joint movement coordinates with scapular motion. In many movements involving the shoulder complex, the humerus and scapula work in concert with one another. For example, after approximately the first 30° of abduction, every 2° of further humeral abduction is accompanied by 1° of scapular rotation. This coordinated action is termed **scapulohumeral rhythm**.

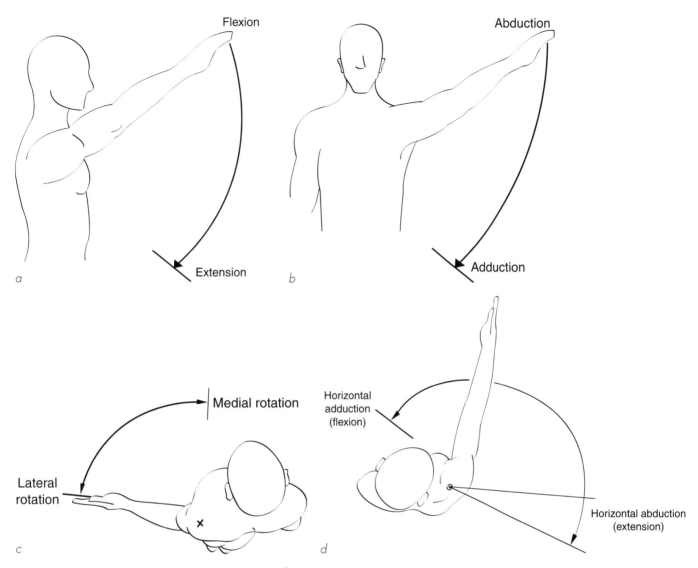

Figure 3.17 Glenohumeral (shoulder) movements: *(a)* flexion and extension; *(b)* abduction and adduction; *(c)* internal (medial) rotation and external (lateral) rotation; *(d)* horizontal abduction and adduction.

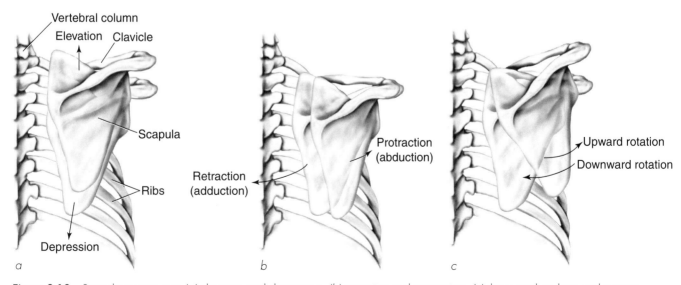

Figure 3.18 Scapula movements: *(a)* elevation and depression, *(b)* retraction and protraction, *(c)* downward and upward rotation.

Movements of the Elbow and Forearm

The elbow complex consists of three joints: humeroulnar, humeroradial, and proximal radioulnar (figure 3.19). The elbow joint proper is formed by the humeroulnar and humeroradial joints. The proximal radioulnar joint works in concert with the distal radioulnar joint to produce movements of the forearm. As a synovial joint, the elbow is surrounded by a thin fibrous capsule that extends continuously from its proximal humeral attachment to the capsule of the proximal radioulnar joint.

The humeroulnar joint is formed by the articulation of the trochlea of the humerus with the trochlear notch of the ulna. The humeroradial joint forms at the junction of the capitulum of the humerus with a shallow depression on the head of the radius.

The proximal radioulnar joint is the articulation between the head of the radius and the ulnar radial notch. At the distal end of the forearm, the ulnar head joins with the ulnar notch of the radius. (Note: The radial head is at the proximal end of the radius, while the ulnar head is at the distal end of the ulna.)

Normal elbow movement is confined to uniplanar flexion and extension at the humeroulnar and humeroradial joints (figure 3.20a). Forearm movements of supination and pronation are produced by conjoint rotations at the proximal and distal radioulnar joints. In anatomical position, the forearm is in a supinated position. From this position, pronation causes the radius to roll over a relatively fixed ulna (figure 3.20b). The reverse occurs in supination when the radius returns to its anatomical position. When the forearm is in a position halfway between full pronation and supination, as when you shake hands, the forearm is in midposition.

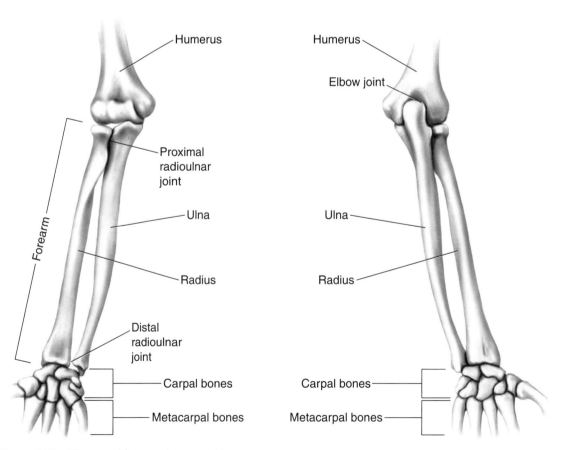

Figure 3.19 Elbow and forearm: bones and joints.

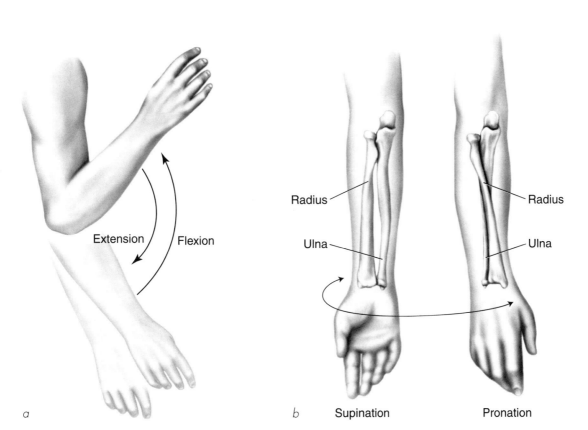

Figure 3.20 Elbow and radioulnar joint movements: *(a)* elbow flexion and extension, *(b)* radioulnar pronation and supination.

Movements of the Wrist and Hand

The wrist (carpus) is not a single joint but rather a group of articulations involving the distal ends of the radius and ulna, along with the carpal bones (figure 2.10, page 32). The wrist complex includes the distal radioulnar, radiocarpal, and intercarpal joints. The hand contains numerous articulations, namely the carpometacarpal (CM), metacarpophalangeal (MP), and interphalangeal (IP) joints. All of these are synovial joints. Structurally, the MP joints are condyloid, while the IP articulations are hinge joints.

The largest joint of the wrist complex is the radiocarpal joint (between the radius and carpals), which allows for flexion and extension (figure 3.21a). The wrist also moves in radial deviation (abduction) and ulnar deviation (adduction) as shown in figure 3.21b.

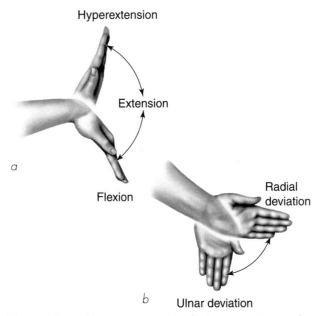

Figure 3.21 Wrist movements: *(a)* flexion, extension, and hyperextension; *(b)* radial and ulnar deviation.

The carpometacarpal (CM) joint of the thumb is a saddle joint that accounts for much of thumb movement. Its structure allows for biplanar movements of flexion, extension, abduction, and adduction (figure 3.22). The second and third CM joints are relatively immovable. The fourth CM joint has limited movement, while the fifth CM joint shows somewhat greater mobility.

Metacarpophalangeal (MP) joints of fingers two through five are synovial condyloid joints that allow the fingers to move in flexion, extension, abduction, and adduction. The thumb's MP joint is a condyloid articulation that permits flexion and limited extension, abduction, and adduction.

Each finger has two interphalangeal (IP) joints (proximal and distal) whose hinge structure permits flexion and extension. The thumb has a single IP joint between its two phalanges.

Movements of the Head, Neck, and Spine

The head is made up of the skull, the brain, and its associated structural components. Structures in the head are protected by an intricate collection of 22 bones. The brain and its protective covering are contained in the cranium, composed of 8 bones: frontal, occipital, ethmoid, and sphenoid bones, along with paired temporal and parietal bones. The anterior and anterolateral aspects of the head are formed by 14 facial bones.

The cranial bones are connected by tight suture joints that are immovable, in the same way as interlocking pieces of a jigsaw puzzle. Normally, the only movable joint in the head is the mandible (jawbone) at the temporomandibular (TMJ) joint formed by the articulation of the mandibular ramus with the temporal bone. The TMJ is a synovial condyloid joint, with joint surfaces separated by an articular fibrocartilage disc.

The head rests on the shoulders atop the most superior portion of the spine. The vertebral column (spine) is a group of 26 vertebrae extending from the base of the skull to its inferior

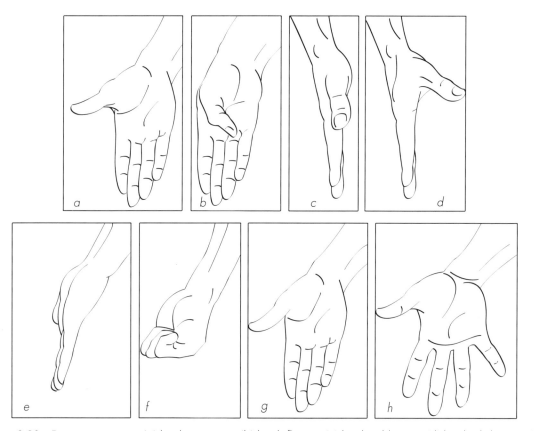

Figure 3.22 Finger movements: (a) thumb extension, (b) thumb flexion, (c) thumb adduction, (d) thumb abduction, (e) finger extension, (f) finger flexion, (g) finger adduction, (h) finger abduction.

termination at the coccyx (tailbone). The spine is divided into five regions (figure 2.11, page 33): cervical (7 vertebrae), thoracic (12), lumbar (5), sacral (1), and coccygeal (1). The sacral and coccygeal vertebrae, while each considered as a single bone, are formed by 5 and 4 fused vertebrae, respectively. Vertebrae in the cervical, thoracic, and lumbar regions are separated by **intervertebral** (IV) **discs** composed of a gelatin-like inner mass (**nucleus pulposus**) surrounded by a layered fibrocartilage network (**annulus fibrosus**) as shown in figure 3.23a. Adjacent vertebrae and the intervening IV disc form a **motion segment** (figure 3.23b). These articulations form symphysis joints and do not have a synovial joint capsule.

At the apex of the spine, the occipital bone of the cranium articulates with the first cervical (C1) vertebra (atlas). These atlantooccipital joints are formed between the concave superior surfaces of the atlas and the convex occipital condyles of the skull.

The joint between the atlas (C1) and the axis (C2) has a unique structure as shown in figure 3.24. A toothlike process (dens) projects superiorly from the axis to articulate with the anterior

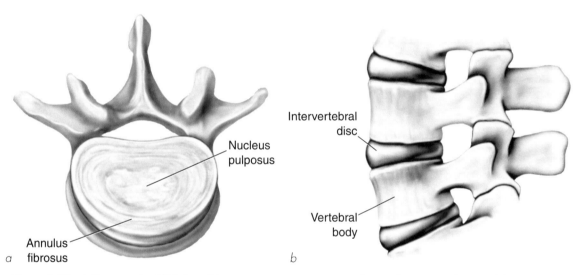

a

Nucleus pulposus

Annulus fibrosus

Intervertebral disc

Vertebral body

b

Figure 3.23   *(a)* Intervertebral (IV) disc; *(b)* motion segment, consisting of two adjacent vertebrae and the intervening IV disc.

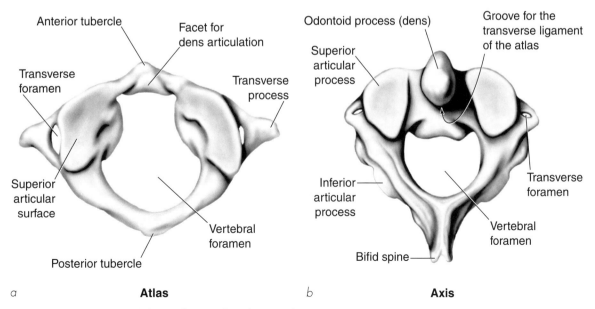

Anterior tubercle

Facet for dens articulation

Transverse foramen

Transverse process

Superior articular surface

Vertebral foramen

Posterior tubercle

a   **Atlas**

Odontoid process (dens)

Groove for the transverse ligament of the atlas

Superior articular process

Inferior articular process

Transverse foramen

Vertebral foramen

Bifid spine

b   **Axis**

Figure 3.24   *(a)* Atlas and *(b)* axis forming the atlantoaxial joint.

arch of the atlas to form the atlantoaxial joint. Synovial joints are found between the articular processes (**zygapophyses**) of adjacent vertebrae. The flatness of adjoining surfaces at these joints permits limited gliding between the segments.

The atlantooccipital joint allows for about 60% of flexion and extension of the head. Head rotation occurs mostly at the atlantoaxial joint, where the unique articulation between the dens of the axis and the atlas allows considerable turning of the head from side to side (figure 3.25a).

As triplanar articulations, the IV joints can move in flexion and extension (sagittal plane), lateral flexion (frontal plane), and rotation (transverse plane) (figure 3.25b). In general, each motion segment has a relatively limited range of motion, depending on the spinal region. Overall spinal movement represents the sum of all motion segment movements.

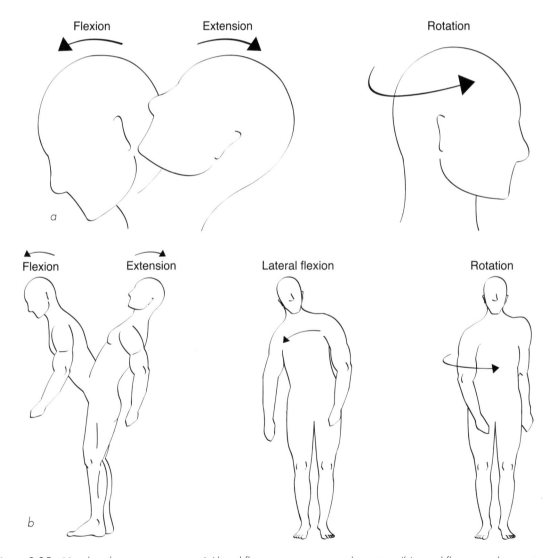

Figure 3.25   Head and spine movements: *(a)* head flexion, extension, and rotation; *(b)* spinal flexion and extension, lateral flexion, and rotation.

Spinal Deformities

As described earlier, the spine forms normal curvatures that help the body accept compressive loads. Injury, disease, and congenital predisposition can cause deformities of the spinal column, leading to abnormal structural alignment or alteration of spinal curvatures. These deformities often result in altered force distribution patterns and pathological tissue adaptations that may lead to or exacerbate other musculoskeletal injuries. There are three primary types of spinal deformity: scoliosis, kyphosis, and lordosis. These deformities are classified by the magnitude, location, direction, and cause and can occur in isolation or in combination. Spinal deformities have long been associated with cardiopulmonary dysfunction. Hippocrates, for example, noted that hunchbacks had difficulty breathing and that patients with scoliosis commonly exhibited dyspnea, or shortness of breath (Padman, 1995).

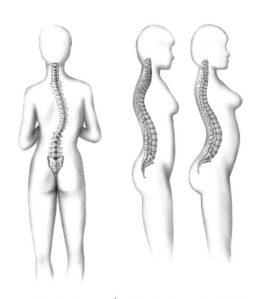

a Scoliosis   b Kyphosis c Lordosis

Figure 3.26 Spinal deformities: *(a)* scoliosis, *(b)* kyphosis, *(c)* lordosis.

Scoliosis is defined as a lateral (frontal plane) curvature of the spine, which is also usually associated with some twisting of the spine (figure 3.26a). Mild spinal deviations are well tolerated and usually asymptomatic (i.e., present no symptoms). Severe deformities, in contrast, can markedly compromise cardiopulmonary function and upper and lower limb mechanics.

Treatment options for scoliosis are either nonoperative or operative. Nonoperative interventions include electrical stimulation, biofeedback, manipulation, muscle strengthening, and bracing. Of these, bracing has proved most successful. For severe scoliotic curvatures, the preferred treatment is operative vertebral fusion in which adjacent vertebrae are fused to prevent further progression of the deformity.

Kyphosis is a sagittal-plane spinal deformity, usually in the thoracic region, characterized by excessive flexion that produces a hunchback posture (figure 3.26b). Kyphosis is more severe in women than in men and is more prevalent with advancing age in both genders. Elderly postmenopausal women are at particular risk, largely because of the strong association between kyphosis and osteoporosis. (Note: In general, *kyphosis* refers to a forward curvature that is concave anteriorly. The thoracic and sacral regions have a natural kyphosis. Thus, hunchback cases actually involve a *hyper*kyphosis, or exaggerated kyphotic curvature. Clinically, the term *kyphosis* often is used, as here, to describe this hyperkyphotic condition.)

The best treatment for kyphosis may lie in prevention. Women with satisfactory exercise habits have a significantly lower incidence of kyphosis, suggesting that physical conditioning programs aimed at proper postural maintenance may delay or prevent the onset of kyphosis associated with aging (Cutler, Friedmann, & Genovese-Stone, 1993).

Clinically, **lordosis** is an abnormal extension deformity, usually in the lumbar region, that produces a hollow or swayback posture (figure 3.26c). (Note: In general, *lordosis* refers to a backward curvature that is concave posteriorly. The cervical and lumbar regions have a natural lordosis. Thus, swayback cases actually involve a *hyper*lordosis, or exaggerated lordotic curvature. Clinically, the term *lordosis* often is used, as here, to describe this hyperlordotic condition.)

This increase in lordotic curvature is commonly seen during pregnancy and results from postural compensations to weight gain, weight distribution (i.e., more weight in the abdominal area), and hormone-induced increases in laxity of connective tissue. Postural changes place unaccustomed stresses on anatomical structures and result in pain and discomfort.

Forward pelvic tilt accentuates lumbar lordosis, which increases the sliding, or shearing, loads on the intervertebral discs and surrounding structures.

concluding comments

Each joint in the body has its own movement potential. Some articulations, such as the glenohumeral joint, possess tremendous freedom and range of movement. Others, such as the tibiofibular joints, have limited movement potential. The body's neuromuscular system considers the movement potential at each joint under its control, coordinates the action of joints, and thereby determines our ability to execute purposeful movements, ranging from simple tasks of daily living to the intricate movements of skilled performers.

critical thinking questions

1. Joints in the human body can be classified as fibrous, cartilaginous, or synovial. Describe the different types (subcategories) of each joint. Include specific examples, and when appropriate, list the specific joint motions and the planes in which they occur (when referenced to anatomical position).

2. What is the function of synovial fluid, and what is believed to cause the cracking (popping) in synovial joints?

3. Define and describe the functional significance of bursae sacs and tendon sheaths.

4. Define arthritis. Give a brief description comparing osteoarthritis, rheumatoid arthritis, and gouty arthritis.

5. What does it mean to be double jointed?

6. The knee and sternoclavicular joints contain fibrocartilage discs, or menisci. Briefly describe their functional significance.

7. Although the knee is often defined as a synovial hinge joint, or even a modified hinge joint, briefly explain why its classification as a bicondyloid joint may be more accurate with respect to both its anatomy and movement capabilities.

8. Describe the primary and secondary curves in the vertebral column, and briefly discuss why they are important. In addition, describe the abnormal curves associated with our vertebral column and some of the anatomical complications they can produce.

suggested readings

Alexander, R.M. (1994). *Bones: The unity of form and function.* New York: Macmillan.

Jenkins, D.B. (1998). *Hollinshead's functional anatomy of the limbs and back* (7th ed.). Philadelphia: Saunders.

Levangie, P.K., & Norkin, C.C. (2001). *Joint structure and function: A comprehensive analysis* (3rd ed.). Philadelphia: Davis.

MacKinnon, P., & Morris, J. (1994). *Oxford textbook of functional anatomy* (Vol. 1, rev. ed.). Oxford, UK: Oxford University Press.

Neumann, D.A. (2002). *Kinesiology of the musculoskeletal system.* St. Louis: Mosby.

Nordin, M., & Frankel, V.H. (2001). *Basic biomechanics of the musculoskeletal system* (3rd ed.). Philadelphia: Lippincott Williams & Wilkins.

four

Myology and the Muscular System

objectives

After studying this chapter, you will be able to do the following:

- Describe the structure and functions of skeletal muscle

- Describe the types of muscle contraction

- Explain the steps in muscle contraction

- Identify muscle fiber types and muscle fiber arrangement

- Explain the length–tension and force–velocity relationships of muscle

- Describe the stretch–shorten cycle and its functional implications

- Explain the process of muscle hypertrophy

- Describe muscle injury and its consequences

Bone may hold man upright, but it is muscle that puts him there. . . . Without bone we couldn't stand, but we could still move; we could—and would—invent boneless sports that we could play, slithering over the ground in some unimaginable fashion. With bone but no muscle, we would be reduced to clacking in the wind.

—John Jerome,
The Sweet Spot in Time

Of all the body's tissues, muscle is unique in its ability to generate force. To understand muscle's role in movement, we first need to understand the fundamentals of muscle structure and function. This chapter focuses on the structure and function of skeletal muscle, including the physiology and mechanics of muscle action, factors affecting muscle force production, neural control of muscle action, and muscle adaptations.

Functions of Skeletal Muscle

Skeletal muscle, which accounts for 40 to 45% of our body weight, plays several important roles in the overall functioning of the human body. In the current context, muscle's most essential role is to produce the forces necessary for human movement, from the basic activities of daily living to the extremes of athletic performance. Muscular actions are largely under voluntary control but may also be elicited through reflex action (e.g., pulling away rapidly from a painful stimulus) or seen in nonreflex movements such as walking that require little, if any, voluntary control. Movements we are so accustomed to performing that they become automatic (e.g., walking, breathing, chewing, coughing, reaching for objects) are sometimes referred to as stereotypical movements.

All muscle tissue is distinguished by four functional properties, or characteristics:

- Excitability
- Contractility
- Extensibility
- Elasticity

Absence or compromise of any (or all) of these properties limits a muscle's ability to function effectively.

Excitability, also known as **irritability**, describes the ability of muscle to respond to a stimulus. The stimulus for skeletal muscles typically comes from the nervous system. As will be described in greater detail in a later section, a muscle's fibers are stimulated by a wave of stimulation conducted along the muscle's length. This conductivity is an important feature of muscle's force production capability.

Contractility, also known as *activity* or *action,* refers to muscle's ability to generate a pulling, or tension, force. (Note: *Contraction* is used by some authors to refer to a muscle's ability to shorten, or change its length.)

Extensibility describes the ability of muscle to lengthen, or stretch, and as a consequence to generate force over a range of lengths. For example, when you shorten the biceps brachii, thereby flexing your elbow, the muscle's extensibility allows it to subsequently lengthen as the elbow extends.

Elasticity refers to a tissue's ability to return to its original length and shape after an applied force is removed. When a muscle and its associated connective tissues are stretched by an external pulling force, for example, its elastic properties allow it to return to its unloaded length once the force is removed.

Structure of Skeletal Muscle

Skeletal muscles are composed of structures that can produce force (**contractile component**) and connective tissues that cannot produce force (**noncontractile component**) but are nonetheless important for the muscle's physiological and mechanical performance. Since muscle cells (fibers) are delicate and can be easily damaged, the collagenous connective tissues around and within the muscle protect the cells from damage. From a mechanical perspective, the connective tissues play a role in accepting and transmitting forces within the muscle and to connected structures.

Muscle's structural hierarchy proceeds from the whole skeletal muscle down to the myofilament level, where active force production actually occurs. We move through these structural levels by first exploring macroscopic (gross) muscle structure and then proceeding to microscopic structures.

Skeletal Muscle Anatomy and Muscle Contraction

Skeletal muscles contain three connective tissue layers that protect and support the muscle, help give it its shape, and contribute collagen fibers to form the tendons at both ends of the muscle. The collagen fibers of the tendons then pass into the bone matrix, enabling the contracting muscle to produce joint movement. The outer connective tissue layer, which surrounds the whole muscle and separates the muscle from surrounding tissues, is the **epimysium**. Within each muscle, bundles of muscle fibers called **fascicles** are separated from one another by **perimysium**. Continuous with the perimysium, and forming the innermost layer of connective tissue, is the **endomysium** that surrounds each muscle fiber (figure 4.1).

A plasma membrane, called the **sarcolemma**, surrounds each muscle fiber and forms narrow tubes, called **transverse tubules**, or **T-tubules**, that pass through the muscle fiber (figure 4.2). These T-tubules are critical because they form channels that allow the electrical signal, needed to stimulate muscle contraction, to pass through the fiber.

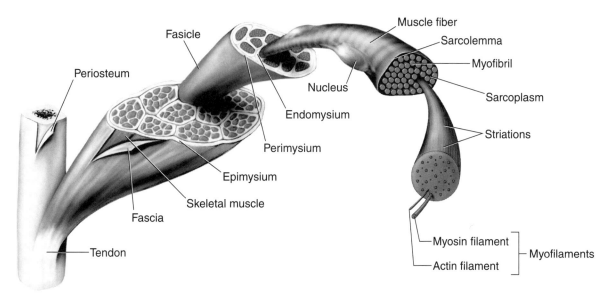

Figure 4.1   Structure of skeletal muscle.

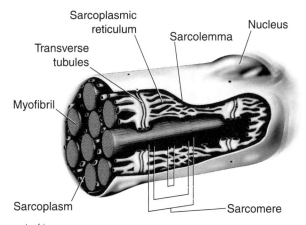

Figure 4.2   Structure of a muscle fiber.

At a microscopic level, each muscle fiber consists of myofibrils, and each myofibril is composed of structural, regulatory, and contractile proteins organized into discrete units called sarcomeres. Each myofibril is surrounded by a fluid-filled system of membranous sacs called the **sarcoplasmic reticulum**. On either side of the T-tubules, expanded ends of the sarcoplasmic reticulum, called terminal cisternae, store the calcium needed for muscle contraction. The **sarcomere** makes up the functional contractile unit of each myofibril (figure 4.3). Each sarcomere is bounded by Z discs (Z lines) and contains actin and myosin myofilaments. How small is a sarcomere? A myofibril 10 mm long contains approximately 4,000 sarcomeres connected end to end. Because of the parallel arrangement of the actin and myosin filaments and the alignment of myofibrils and their respective sarcomeres, skeletal muscles have a striped, or striated, appearance.

Sarcomeres contain three kinds of proteins: (1) contractile proteins (actin and myosin) that produce force during contraction, (2) regulatory proteins (tropomyosin and troponin) that help turn the contractile process on and off, and (3) structural proteins (e.g., titin, myomesin) that maintain alignment of the actin and myosin filaments.

The contractile proteins consist of **myosin** and **actin** filaments, otherwise known as thick and thin filaments, respectively. The tail regions of all the myosin molecules connect at the M line, whereas the myosin heads are all located closer to the Z lines and are located between the actin filaments (figure 4.3).

Every actin filament is anchored to a Z line and extends toward the middle, or M line, of the sarcomere. The actin filaments also contain two regulatory proteins known as **tropomyosin** and **troponin**. Tropomyosin molecules bonded together form continuous strands that extend the length of the actin filament, while troponin molecules are located at the junction of each successive pair of tropomyosin molecules. When the muscle is relaxed, these tropomyosin strands cover the myosin-head binding sites on the actin filament.

In addition to the contractile and regulatory proteins just discussed, each sarcomere contains structural proteins that contribute to the alignment, stability, elasticity, and extensibility of the myofibril. Two prominent structural proteins are titin and myomesin. **Titin** stabilizes the alignment of the thick filaments by connecting each myosin filament to both the Z disc and the M line. The titin that extends from the Z disc to the beginning of the thick filament is believed to help keep the myosin filaments centered between the Z discs, and it also accounts

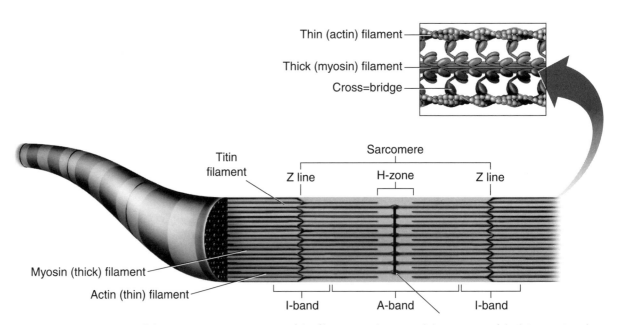

Figure 4.3 Structure of the sarcomere, arrangement of the filaments within it, and the structure of thick (myosin) and thin (actin) filaments.

for much of the elasticity and extensibility of myofibrils. Titin therefore may help the sarcomere return to its resting length after the muscle has contracted or been stretched. **Myomesin** molecules form the M line that anchors the myosin filaments, and the titin filaments, in the center of the sarcomere.

Types of Muscle Contraction

Muscle **contraction**, or **action**, is the internal state in which a muscle actively exerts a force, regardless of whether it shortens or lengthens. There are three types of muscle contraction:

1. **Concentric** (shortening): The turning, or rotational, effect (torque) produced by the muscle at a given joint is greater than the external torque (created by an external force such as a held weight), and therefore, the muscle is able to shorten while overcoming the external load. For example, concentric muscle action is used to flex the elbow from a fully extended position.

2. **Isometric** (*iso* = same; *metric* = length): The torque produced by the muscle is equal and opposite to the external torque, and therefore, there is no limb movement. (See box below.) For example, isometric muscle action is used to hold the elbow at 90° of flexion.

3. **Eccentric** (lengthening): The torque produced by the muscle is less than the external torque, but the torque produced by the muscle causes the joint movement to occur more *slowly* than the external torque, acting by itself, would tend to make the limb move. For example, eccentric muscle action is used to slowly extend the elbow from a flexed position.

How do you determine which type of muscle contraction is occurring during exercise? Perhaps the easiest application is weightlifting. For any exercise (e.g., bench press, squat, shoulder press, elbow curl, seated row), determine when during the lift you are overcoming the weight (i.e., overcoming gravity). That is the concentric phase. In contrast, when you let the weight overcome the force your muscles are producing (i.e., when the weight is moving with gravity), you are performing the eccentric phase. If at any time during the movement you stop and hold the weight in a fixed position, you are performing an isometric phase. For example, in the flat bench press, lowering the weight to your chest is the eccentric phase, and pressing the weight off the chest is the concentric phase. If at any time during the bench press you stop and hold the weight, then you are performing an isometric phase.

In addition to the three contraction types just listed, two important terms are often used to define properties of a contraction. The term **isokinetic** describes constant angular velocity about a joint. It is possible, therefore, to produce an isokinetic concentric and an isokinetic eccentric action. Isokinetic contractions are commonly used in experiments that test the relative effectiveness of different exercise devices for recruiting specific muscle groups. By using isokinetic contractions, the experimenters can control for the effects of contraction velocity on muscle force production.

Isometric: Is It Really?

Derived from *iso*, meaning same, and *metric*, meaning measure or length, the term *isometric* typically is used to describe tasks in which muscle force is produced with no resulting movement. Holding a weight in a fixed position, for example, requires so-called isometric action. It is often inferred that the absence of joint movement means the muscle is also acting isometrically, or not changing its length. In actuality, when a joint is held motionless while muscle force is generated, the entire **musculotendinous unit** is isometric. Within the musculotendinous unit, however, the active muscle shortens slightly and pulls to lengthen the tendon. Slight shortening of the muscle combined with a bit of tendon lengthening results in no net length change (isometric) in the musculotendinous unit. The length changes in the muscle and tendon are minimal, of course, but to say that constant, or unchanging, joint angles are associated with true isometric muscle action is technically incorrect.

Isotonic literally means constant tension. This condition does not occur in intact human subjects (i.e., *in vivo*) because the level of muscle force varies continuously and rarely, if ever, is constant throughout a movement. Isotonic conditions are practical only in isolated muscle preparations in laboratory experiments. A more accurate term for human actions is **isoinertial**, which means constant resistance. For example, when performing an elbow curl with a 25 lb dumbbell, you are performing an isoinertial action throughout the movement because the external resistance remains constant.

Excitation–Contraction Coupling

The physiological steps that produce a muscle contraction involve passage of an electrical signal (**action potential**) through the sarcolemma and the eventual swiveling (pivoting) of the myosin heads to produce force. This process, called **excitation–contraction coupling**, is outlined as follows:

1. The electrical signal passes along the axon of the **lower motor neuron** (LMN) to the presynaptic membrane of the terminal ending, causing calcium channels to open.

2. Calcium from the surrounding fluid flows into the terminal ending and facilitates the movement of the synaptic vesicles to the presynaptic membrane. Each vesicle contains the neurotransmitter **acetylcholine** (ACh).

3. The synaptic vesicles bind to the presynaptic membrane of the terminal ending and release acetylcholine into the synaptic cleft. This process is called **exocytosis**.

4. Acetylcholine crosses the synaptic cleft and binds to ACh receptors on the sarcolemma. This binding causes the sodium channels to open, and if enough sodium rushes into the muscle fiber, an action potential will pass across the sarcolemma in all directions.

5. The electrical signal (action potential) then passes along the sarcolemma and into the fiber through the T-tubules. The T-tubules pass through the muscle fiber and around all of the myofibrils.

6. As the electrical signal passes through the T-tubules, it causes the release of calcium from the sarcoplasmic reticulum.

7. The calcium then binds to troponin. This interaction between the troponin and calcium is believed to cause a shift of the tropomyosin strands off the myosin-head binding sites on the actin filament.

8. The myosin heads bind to the actin filament and swivel (pivot) toward the M line. A new energy molecule, **adenosine triphosphate** (ATP), must be added to the myosin head for it to detach from the actin. This cycling of the myosin heads continues as long as the neural signal is maintained, calcium stays bound to troponin, and the ATP needed to supply the energy for contraction is replenished.

9. The muscle will stop contracting, however, when the stimulation stops and acetylcholine is no longer released from the terminal ending of the LMN.

10. If the signal stops, calcium will then be actively pumped back into the sarcoplasmic reticulum, and the troponin–tropomyosin complex will settle back into its original position on the actin filaments, blocking the myosin-head binding sites. With no more interaction between the actin and myosin filaments, the muscle relaxes and returns to its resting, or starting, length.

The mechanical sum of the 10 steps just described makes up the **sliding filament model** of muscle contraction. The sliding filament model is generally accepted as the best description of the steps that produce muscle contraction.

How many times must the myosin heads swivel to produce a muscle contraction? The range of joint motion determines how many cycles each myosin head needs to perform. In other words, during shortening or lengthening muscle actions, each myosin head undergoes many repeated but independent cycles of asynchronous movement. During an isometric contraction, the myosin heads still bind to the actin and swivel to produce force, but unlike a concentric contraction, this cycle is not repeated to shorten the muscle. During an eccentric contraction, the external load, or body weight, overcomes the normal tendency for the myosin heads to swivel toward the M line. The actin–myosin coupling is still forming, but the bond between the proteins is broken as the myosin heads are forcefully pulled toward the Z line.

When performing overload training with a weight that exceeds your concentric limit, the muscle has no choice but to perform an eccentric contraction. Eccentric actions, particularly at this intensity, produce the most muscle damage and therefore stimulate the most regeneration and growth. This explains why high-intensity eccentric contractions, referred to as *forced negatives*, are popular with bodybuilders seeking to maximize muscle size.

Muscle Fiber Types

Human skeletal muscles are a composite of different fiber types, with the percentage of fiber type varying from muscle to muscle and person to person. Based on their mechanical and contractile properties, skeletal muscles are divided into two distinct fiber types: slow twitch (ST) and fast twitch (FT). Metabolically, muscle fibers can be classified as **oxidative** or **glycolytic**. Because of their different metabolic profiles, slow-twitch fibers are classified as slow oxidative (SO) fibers, and fast-twitch fibers are further divided into two principal subcategories: fast oxidative glycolytic (FOG) fibers and fast glycolytic (FG) fibers. SO fibers have the highest number of mitochondria, oxidative enzymes, myoglobin, and capillaries and therefore generate ATP primarily through aerobic (i.e., using oxygen) pathways.

In contrast, FG fibers have relatively few mitochondria, oxidative enzymes, and capillaries and a low myoglobin content. They do store large amounts of glycogen, however, and generate ATP mainly through anaerobic glycolysis (the metabolism of sugar without using oxygen). Because FG fibers contain the most myofibrils and use ATP quickly, they are the strongest and fastest fibers. In other words, FG fibers are designed for intense, short-duration anaerobic movements. As indicated in table 4.1, FOG fibers are often referred to as intermediate fibers because they share characteristics of both SO and FG fibers. Some of the contractile, anatomical, and metabolic characteristics that distinguish SO, FOG, and FG fibers are listed in table 4.1.

Table 4.1   Names and Properties of Skeletal Muscle Fiber Types

Property	SO	FOG	FG
Color	Red	White/red	White
Myoglobin content	High	Intermediate/high	Low
Capillary density	High	Intermediate	Low
Oxidative enzyme content	High	Intermediate/high	Low
Mitochondrial content	High	Intermediate	Low
Stored lipids	High	Intermediate/high	Low
Fatigue resistance	High	Intermediate	Low
Glycogen content	Low	Intermediate	High
Glycolytic enzyme content	Low	Intermediate/high	High
Myosin ATPase activity	Low	High	High
Fiber diameter	Small	Intermediate	Large

Alternative names of fibers:

Slow twitch = Type I = slow oxidative (SO)

Fast twitch = Type IIa = fast fatigue resistant (FR) = fast oxidative glycolytic (FOG)

Fast twitch = Type IIb = fast fatigable (FF) = fast glycolytic (FG)

Is it possible to change fiber types through training? Training may not change the neural control (i.e., the specific neuron and therefore its size, threshold, and speed of conduction velocity), but it can change the metabolic profile of the fibers. For example, endurance training can increase the oxidative capacity of all three fiber types, so with intense endurance training, what started off as an FG fiber metabolically may take on the metabolic profile of an FOG fiber. Although neurologically FG fibers are still the fastest and have the highest myosin ATPase activity (the enzyme used to break down ATP on the myosin molecule), a great deal of movement speed comes from the proper training of the nervous system. A well-trained primarily slow-twitch athlete can still be fast. The genetic advantage in speed and power, however, still goes to the predominantly fast-twitch individual. Likewise, the genetic advantage for endurance still goes to the individual with a predominance of SO fibers.

Muscle Fiber Recruitment

The neuromuscular junction represents the synaptic connection between the lower motor neuron (LMN) and the sarcolemma. A **motor unit** is defined as a single LMN plus all the muscle fibers it innervates or has **innervation** with. Because a lower motor neuron innervates only one muscle fiber type, the SO, FOG, and FG classification applies not only to muscle fibers but also to motor units. Muscle contraction is the result of many motor units firing asynchronously and repeatedly. When exercise progresses from low to high intensity, SO fibers will be recruited first, followed by FOG and then FG fibers (figure 4.4). The SO fibers produce the least amount of force because they contain the fewest actin and myosin filaments. On the other end of the spectrum are the FG fibers, with the highest number of actin and myosin filaments and the highest myosin ATPase activity.

When FG fibers are stimulated, therefore, they produce the most force and do so the most quickly. When you progress from low-to high-intensity exercise, your nervous system recruits your muscle fibers in order from the most aerobic (the most resistant to fatigue) to the least aerobic (the easiest to fatigue). However, if you choose to produce a great deal of force or power quickly or explosively, your nervous system will override this normal aerobically energy-efficient hierarchy by firing all three fiber types at the same time. In this case, there would actually be a preferential recruitment of fast-twitch fibers.

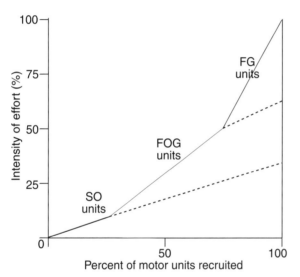

Figure 4.4   Recruitment order of slow oxidative (SO), fast oxidative glycolytic (FOG), and fast glycolytic (FG) muscle fibers.

Muscle Fiber Arrangement

Muscle fiber arrangement within our skeletal muscles falls into one of seven different categories (figure 4.5):

    fusiform, or longitudinal

    unipennate

    bipennate

    multipennate

    triangular, or radiate

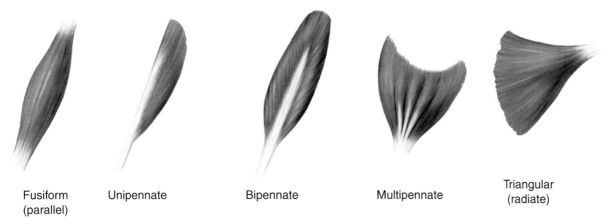

| Fusiform (parallel) | Unipennate | Bipennate | Multipennate | Triangular (radiate) |

Figure 4.5 Muscle fiber arrangements.

To determine the fiber arrangement of a muscle, on a sketch of the muscle simply draw a line from the muscle's origin through its insertion. If the muscle fibers run parallel with that line, the muscle is fusiform (e.g., biceps brachii, semitendinosus), longitudinal, or quadrate. In pennate muscles, fibers are oriented at oblique angles (normally <30 degrees) to the tendon's line of pull. A unipennate muscle has one set of fibers, all with the same line of pull (e.g., semimembranosus); a bipennate muscle has two sets of fibers with different angles (e.g., rectus femoris); and a multipennate muscle has many sets of fibers at a variety of angles (e.g., deltoid).

The advantage of **pennation** is that for any given volume, more muscle fibers function in parallel, giving the muscle a greater functional cross-sectional area. In other words, pennation is the body's way of packing a greater number of muscle fibers into a smaller volume, therefore producing a muscle designed more for force production than for speed of contraction.

Although fusiform muscles can produce considerable force, they are designed to maximize speed of contraction. For example, given a pennate and fusiform muscle of equal length and volume, the fusiform muscle will have more sarcomeres in series because the muscle fibers are in line with the tendon. If both muscles are stimulated and all their sarcomeres shorten the same distance, then the muscle with the greater number of sarcomeres in series will shorten through a greater distance in the same period of time.

In muscles classified as triangular (also called radiate or fan shaped), fibers fan out from a relatively large origin to a relatively small insertion (e.g., pectoralis major, latissimus dorsi). This design combines high force production with high speed of contraction by packing numerous sarcomeres both in parallel and in series.

Although training cannot change a muscle's architecture, understanding design differences can help us recognize a muscle's potential for injury. For example, the quadriceps are designed for force production, whereas the hamstrings are designed for rapid shortening. Because of these design differences and the fact that the hamstring muscles (semitendinosus, semimembranosus, and biceps femoris long head) cross both the hip and the knee, these muscles are more susceptible to tearing than are the quadriceps in explosive, high-power events such as sprinting.

Length–Tension Relationship

We have known since the late 1800s that the length of a muscle affects its isometric force production capability (Lieber, 2002). Earlier in this chapter we discussed how the nervous system stimulates muscle fibers to produce force. What wasn't discussed, however, is that a muscle's force production is also affected by its passive elements (e.g., titin strands, tendons, and connective tissue sheaths) that are not under neural control.

A muscle's force production capability as a function of its length is shown in figure 4.6. The active component represents the force generating capacity of the sarcomeres, and therefore, a

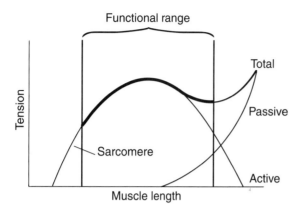

Figure 4.6 Length–tension relationship of skeletal muscle.

Adapted from Aidley 1978.

myofibril. If a sarcomere is too short, there will be complete overlap between the actin filaments, the myosin filaments will press against the Z lines, the myosin heads will be unable to bind to the actin, and the sarcomere will be incapable of producing force. As the sarcomere is lengthened, it will reach its optimum force production capability when there is maximum overlap (and binding capability) between the myosin heads and the actin molecules. If the sarcomere is stretched too much, however, there won't be any actin and myosin overlap, and force production will drop to zero.

As the sarcomere is lengthened beyond its optimal force potential, the passive elements begin to lengthen and mechanically want to recoil. As the **length–tension relationship** in figure 4.6 shows, the passive tension continues to increase as the muscle is lengthened and helps compensate for the loss in force production by the sarcomere. The muscle's total force production capability, therefore, is calculated by summing the active and passive components.

The data used to generate the length–tension curve (figure 4.6) were originally collected from isolated animal muscle using isometric contractions. How then does the length–tension relationship apply to the muscles in our body? The vertical lines added to figure 4.6 represent the fact that *in vivo* our joints do not allow us to shorten or lengthen a muscle to such extremes that all our sarcomeres would be incapable of producing force. Please note, the exact location of these vertical lines are only approximations and will vary depending on the muscle being analyzed. Figure 4.6 shows that when a muscle is maximally shortened, its force production capability is due solely to its active component, but when the same muscle is maximally lengthened, its force production now depends on both its active and passive components.

The length–tension relationship of muscle is easier to understand in application. For example, can you lift more weight in a standing versus a seated calf-raise? Can you leg curl more weight when your pelvis is anteriorly tilted than when it is flat on the bench? The answer to both questions is yes. Why? In each case you are changing a joint position to pre-lengthen one or several of the principal muscles being trained. Because the gastrocnemius crosses both the ankle and the knee, the extended knee lengthens the muscle and increases its force production capability when performing a standing calf raise. In a seated calf raise, the gastrocnemius is too short to maximize force production, making this an excellent exercise to target the soleus, which crosses only the ankle joint and whose length is unaffected by knee flexion and extension.

Anteriorly tilting the pelvis, or flexing the hip, in a leg curl exercise lengthens the long head of the biceps femoris, semitendinosus, and semimembranosus, and therefore increases their force production capability during the lift. During the leg curl, the ankle is usually in anatomical position or slightly dorsiflexed. Because the gastrocnemius both flexes the knee and plantar flexes the ankle, dorsiflexion lengthens the muscle and increases its force production capability. A similar action happens in cycling, where anterior pelvic tilt is not only used to improve aerodynamics, but also to increase the force output of the hamstrings and gluteus maximus by increasing their length.

Force–Velocity Relationship

The force production capability of skeletal muscle depends, in part, on the contraction velocity. This dependency is known as the **force–velocity relationship.** Each point on the force–velocity diagram (figure 4.7) represents the maximum force for any given contraction velocity when the muscle is maximally stimulated. According to the diagram, the faster a muscle contracts concentrically the less force it can produce. To understand why this is true we need to consider what happens to the myofibril during a contraction. As the shortening velocity increases, the number of myosin heads binding to actin decreases, thereby decreasing the force produced by the muscle. When the muscle is maximally stimulated, more force is produced isometrically than can be developed at any speed of concentric action, and as figure 4.7 shows, the greatest force can be produced in an eccentric action. Because the curve on the force–velocity diagram represents maximal stimulation of the muscle, the muscle functions submaximally underneath the curve in most everyday tasks.

To experience the force–velocity relationship, try the following experiment. Perform a series of dumbbell elbow curls. Execute each repetition with maximum effort (speed), using the same range of motion and proper form. Starting with no dumbbell, then using dumbbells that increase in 5 lb increments, perform one maximum-speed curl for each weight, allowing sufficient rest between lifts to reduce fatigue effects. Rate the speed of contraction for each lift using subjective terms such as *very fast, fast, moderate, slow,* and so on. What happens to movement speed as you increase the weight? It slows down. Once you reach a weight you can't lift through the full range of motion, you have reached or are very close to your isometric limit. The weight you are holding in that static (stationary) position is heavier than any weight you can lift concentrically.

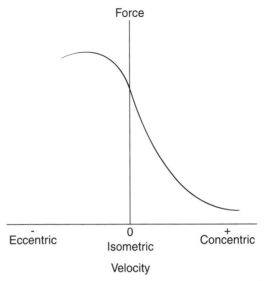

Figure 4.7 Force–velocity relationship of skeletal muscle.

For illustration purposes, assume your isometric limit for the dumbbell elbow curl is 50 lb. There would be no point in using a 55 lb dumbbell from the starting position because this weight exceeds your concentric limit. But if a spotter hands you the 55 lb dumbbell at the top position (i.e., the most flexed elbow position), the descent of the weight can be controlled using eccentric muscle action. Notice the movement occurs in the same direction as gravity, but slower than gravity wants the dumbbell to move. With each additional 5 lb increase in dumbbell weight, not only will your speed start increasing, but it will also get harder to control. Soon a weight will be reached that is too heavy to resist, or if you do, you run the risk of injury. You have just proved that when a muscle is maximally stimulated, it can handle a greater load during eccentric actions than during isometric or concentric actions, and the faster the muscle contracts concentrically, the less force it can produce.

Stretch–Shorten Cycle

The **stretch–shorten cycle** may be defined as an eccentric action immediately followed by a concentric action. The benefit of this cycle is that the normal concentric force production capability of the muscle is enhanced. Four mechanisms have been proposed to explain this enhancement: time for force development, elastic energy, force potentiation, and reflexes (Enoka, 2002). These mechanisms—and which combination of them applies to a given movement—remain controversial.

Let's consider one of the mechanisms, elastic energy, in some detail. During the eccentric phase, elastic energy is stored in the titin strands and connective tissue sheaths, then released during the following concentric contraction. The muscle's normal force production is augmented by the addition of the released elastic energy. The added benefit is that the storage and release of the elastic energy did not require any additional ATP. The concentric action needs to immediately follow the eccentric action, otherwise some of the stored energy will be dissipated in the form of heat.

In general, the more ballistic the cycle, the greater the amount of energy stored. If the joint range of motion is too small, stored energy will be minimal. If the range of motion is too great,

Feats of Muscular Strength and Endurance

Something in our nature seems drawn to the limits of human performance. We are fascinated by world records, both traditional and unusual. The limits of human strength can be seen in reports of a frantic parent who lifts an automobile off a trapped child. In more measured settings, limits of strength are determined by records set in weightlifting competitions such as the Olympic Games.

One classic measure of human strength is the deadlift, in which a weight is lifted from the ground to a position with knees locked and body erect. The record for the deadlift, relative to body weight, is held by Lamar Gant, who in 1985 deadlifted 660 lb (300 kg) to become the first person to lift *five times* his own body weight. Gant weighed a mere 131 lb (60 kg) (Cunningham, 2002).

Records continue to be broken, which raises interesting questions such as, "Are there limits to human performance?" and if so, "What are those limits?"

more energy may be stored, but the muscle's leverage may be compromised. For example, consider the technique used to maximize a vertical jump. Normally the jumper takes a few quick steps; unweighs by flexing the hips, knees, and ankles; and then explodes upward into the jumping, or concentric, phase. If the unweighing phase is too deep, the leverage of the hip and knee extensors will be compromised, making it difficult to explode vertically.

Strength and power athletes routinely use the stretch–shorten cycle in their training by performing plyometrics. **Plyometrics** is a form of exercise that uses the force-enhancing characteristics of the stretch–shorten cycle. Walking and running also utilize stretch–shorten cycles. Because of greater joint range of motion and more rapid force generation in running, however, runners typically benefit more from the stretch–shorten cycle than do walkers.

Muscle Hypertrophy

Muscle hypertrophy involves an increase in the size and number of myofibrils, which increases the size of the muscle fibers, the fascicles, and ultimately, the overall muscle belly. In addition, the connective tissue must also expand to accommodate the increasing size of the muscle.

In response to repeated overload stress, new actin and myosin filaments are added to the periphery of the sarcomeres within a myofibril, under the sarcoplasmic reticulum, until the myofibril increases to approximately twice its original diameter. Then, for reasons still not completely understood, the myofibril splits down its longitudinal axis, forming two parallel myofibrils. Studies have also shown that fast-twitch fibers tend to hypertrophy more quickly, and to a greater extent, than slow-twitch fibers. People with a predominance of fast-twitch fibers, therefore, may hypertrophy at a greater rate and achieve more mass than a person endowed with primarily slow-twitch fibers. During hypertrophy, the sarcoplasmic reticulum must increase in size to match the increase in diameter of the myofibrils, and the sarcolemma must grow to match the increase in size of the muscle fiber.

Although isometric, concentric, and eccentric actions can all produce muscle hypertrophy, eccentric contractions provide the greatest stimulus, as explained earlier. When a weight is lowered with gravity in an eccentric action, the nervous system recruits fewer muscle fibers. This produces more muscle and connective tissue damage to the fibers involved and therefore provides a greater stimulus for muscle growth.

Muscle Injury, Pain, and Soreness

Although skeletal muscles are capable of generating high forces without injury, too much force transmitted through the musculotendinous unit can, and often does, produce injury. Musculotendinous injuries are called strain injuries, or **strains**. (Note: The term *strain* also is used to describe the physical deformation of a material, whether or not injury or damage occurs. This *mechanical strain* will be explained in chapter 6.)

Strain injuries can occur as tearing where the tendon inserts into the bone (osteotendinous junction), within the body of the tendon where the tendon joins with the muscle (myotendinous or musculotendinous junction), or within the belly (substance) of the muscle itself. Recent research has shown that the myotendinous junction and the area immediately adjacent to it are common sites of strain injury.

Injury typically occurs during forced lengthening, or eccentric, muscle action employed to control or decelerate high-velocity movements (e.g., sprinting, throwing, jumping) and is hastened by many factors, including fatigue, muscle imbalances, inflexibility, and insufficient warm-up.

Certain muscles seem more prone to strain injuries. The muscles of the hamstring group (on the posterior side of the thigh) are especially susceptible to injury. This may, in part, be due to the fact that these muscles are biarticular (i.e., have action at two joints). This structural arrangement dictates that muscle length is determined by the combined action of the hip and knee joints. Hip flexion and knee extension both cause lengthening of the hamstring muscles. Simultaneous hip flexion and knee extension places the hamstrings in a lengthened state that contributes to the muscles' risk of injury. If these movements are made quickly, as when a sprinter swings her leg forward through the air and plants her foot on the ground, the chance of injury increases considerably. Hamstring injuries of this type often happen at the myotendinous junction. So, a muscle's gross and microscopic anatomical structure, biarticular arrangement, and involvement in controlling high-velocity movement all contribute to its susceptibility to strain injuries.

Muscle pain and soreness felt during and immediately after exercise are normally attributed to tissue swelling (**edema**) and the accumulation of lactate and other metabolites. These substances diffuse out of the muscle and either directly stimulate free nerve endings or cause edema, resulting in pain. Edema in this case is typically the result of fluid from blood plasma moving into the muscle tissue. The good news is that this pain normally disappears within minutes to hours after exercise.

After strenuous exercise, it is not uncommon for an individual to experience muscle soreness. Although muscle soreness may occur more frequently during the initial stages of a new training program, or in someone just starting a program, it can also occur in well-trained athletes if they significantly increase their training volume. We normally do not perceive this **delayed onset muscle soreness** (DOMS) until 24 to 48 hours after the exercise session. Electron micrograph images of skeletal muscles taken after intense exercise show tears in the connective tissue, sarcolemma, sarcoplasmic reticulum, actin and myosin filaments, and Z lines of muscle fibers. DOMS is typically accompanied by edema, tenderness, and stiffness. Although we do not understand all the factors that produce DOMS, muscle damage appears to be the major cause. Because eccentric contractions produce more muscle damage than either concentric or isometric contractions, they are the principal cause of DOMS. Adequate recovery after intense training, therefore, allows the muscle to adapt to the training stress without producing continual soreness. If athletes dramatically increase training volume, however, they may again experience DOMS. Because this training-induced soreness is not indicative of any serious injury or problem, continued training at a reduced intensity is generally recommended. Lighter training speeds the repair process by increasing metabolism, supplying required nutrients, and removing waste products from the sore muscles.

concluding comments

Muscles are the engines that power all of our body's movements. Among muscle's remarkable characteristics are its ability to generate force and to adapt to training. This chapter provides the basics for understanding how human skeletal muscle generates force and what factors affect this force production. We review the fundamentals of muscle structure, from the macroscopic whole-muscle level down to the microscopic events involved in muscle contraction at the actin–myosin level, and discuss how muscle produces force through the excitation–contraction coupling process. We discuss many factors that help modulate force output, including contraction type, fiber type, recruitment, fiber arrangement, muscle length, and velocity.

No other tissue in the human body responds to intense training more noticeably than muscle. An elite bodybuilder, for example, conjures an entirely different image than a world-class marathoner. The size of an athlete's muscles readily lets us know, or at least enables us to guess, whether they have trained for endurance, strength, power, or hypertrophy.

In subsequent chapters, we use this information to examine how muscles work in specific situations to produce everyday movements and how misuse and overuse can result in muscle injury.

critical thinking questions

1. Explain why fast glycolytic fibers are faster and stronger than slow oxidative fibers.
2. Why, *in vivo*, does the length–tension relationship not drop to zero when a muscle is either at its shortest or greatest length?
3. Provide at least two specific weightlifting exercises where variations in body position affect the length–tension relationship of one or more of the principal muscles being trained.
4. When a muscle is maximally stimulated, why does the muscle's force production decline the faster it contracts concentrically?
5. When a muscle is maximally stimulated, why can you handle more weight during an isometric contraction than during a concentric contraction?
6. When a muscle is maximally stimulated, why can you handle a greater load during an eccentric contraction than during either an isometric or concentric contraction?
7. Explain how a stretch–shorten cycle can increase a muscle's normal concentric force production capability.
8. Explain why eccentric contractions provide a greater stimulus for muscle hypertrophy than either isometric or concentric contractions.
9. Explain why muscles that function across more than one joint are more likely to be injured during ballistic movements than muscles that function across only one joint.
10. Of the three contraction types, why are eccentric contractions believed to be the major cause of delayed onset muscle soreness (DOMS)?
11. Describe one or more athletic events that would alter the normal recruitment order of muscle fibers.
12. Why and how does a muscle hypertrophy in response to overload stimulus?

suggested readings

Enoka, R.M. (2002). *Neuromechanics of human movement* (3rd ed.). Champaign, IL: Human Kinetics.

Kreighbaum, E., & Barthels, K.M. (1996). *Biomechanics: A qualitative approach for studying human movement* (4th ed.). Needham Heights, MA: Allyn & Bacon.

Lieber, R.L. (2002). *Skeletal muscle structure, function and plasticity: The physiological basis of rehabilitation* (2nd ed.). Philadelphia: Lippincott Williams & Wilkins.

McArdle, W.D., Katch, F.I., & Katch, V.L. (2001). *Exercise physiology: Energy, nutrition, and human performance* (5th ed.). Baltimore: Lippincott Williams & Wilkins.

Neumann, D.A. (2002). *Kinesiology of the musculoskeletal system: Foundations for physical rehabilitation*. St. Louis: Mosby.

Tortora, G.J., & Grabowski, S.R. (2003). *Principles of anatomy and physiology* (10th ed.). New York: Wiley.

Wilmore, J.H, & Costill, D.L. (2004). *Physiology of sport and exercise* (3rd ed.). Champaign, IL: Human Kinetics.

five

Muscles of Movement

objectives

After studying this chapter, you will be able to do the following:

- Explain how muscles are named
- Determine the functional actions of muscle
- Identify the muscles acting at the major joints of the body

Skeletal muscle represents the classic example of a structure–function relationship. At both the macro- and microscopic levels, skeletal muscle is exquisitely tailored for force generation and movement.

—*Richard L. Lieber,* Skeletal Muscle Structure and Function

n the previous chapter, we examined muscle structure and function with little mention of specific muscles. In this chapter, we take a more detailed look at those skeletal muscles responsible for producing and controlling the movements of the trunk and limbs. Of the approximately 600 muscles in the body, fewer than one third are essential for gross body movements (figure 5.1). We will now examine these muscles in detail.

Muscle Names

Muscles are named in various ways, and often something in the name helps describe the muscle. A muscle's name might describe its size, shape, location, action, or attachment sites; the number of muscle bellies; or the direction of its fibers, as the following examples show:

- Size: The pectoralis major is a large muscle on the anterior pectoral, or chest, region. In addition to *major*, other size-related terms include *maximus* (largest, e.g., gluteus maximus), *minimus* (smallest, e.g., gluteus minimus), *longus* (long, e.g., peroneus longus), and *brevis* (short, e.g., peroneus brevis).

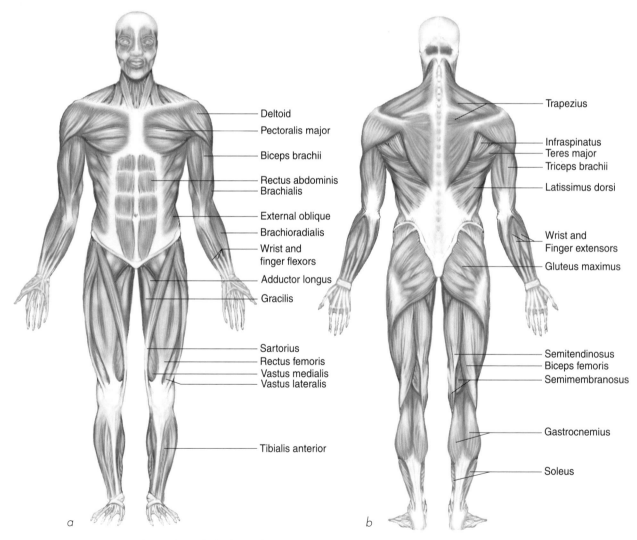

Figure 5.1 Principal superficial skeletal muscles: *(a)* anterior view, *(b)* posterior view.

Reprinted, by permission, from National Strength and Conditioning Association, 2000, *Essentials of strength training and conditioning,* 2nd edition (Champaign, IL: Human Kinetics), 29.

- Shape: The deltoid is a shoulder muscle shaped like a triangle whose name is derived from the triangularly shaped Greek letter delta (Δ). The trapezius is named for its trapezoidal (four-sided) shape.
- Location: The tibialis anterior lies on the anterolateral side of the lower leg, adjacent to the tibia.
- Action: The flexor digitorum muscle flexes the fingers or toes.
- Attachment sites: The sternocleidomastoid attaches from the sternum and clavicle to the mastoid process.
- Number of origins, or muscle bellies: The biceps brachii on the anterior side of the upper arm (brachium) has two heads, or bellies. The triceps brachii has three heads.
- Fiber direction: The rectus abdominis is a muscle whose fibers run parallel along the midline of the trunk's anterior surface (*rectus* = straight). Other directional terms include *transversus* (perpendicular to the body's midline, e.g., transversus abdominis) and *oblique* (diagonal to the midline, e.g., external oblique).

See table 5.1 for a more complete list of muscle terminology.

Table 5.1 Muscle Terminology

Factor	Term	Meaning
Size	Brevis	Short
	Gracilis	Slender
	Lata	Wide
	Latissimus	Widest
	Longissimus	Longest
	Longus	Long
	Magnus	Large
	Major	Larger
	Maximus	Largest
	Minimus	Smallest
	Minor	Smaller
	Vastus	Great
Shape	Deltoid	Triangular
	Orbicularis	Circular
	Pectinate	Comblike
	Piriformis	Pear shaped
	Platys-	Flat
	Pyramidal	Pyramid shaped
	Rhomboideus	Rhomboidal
	Serratus	Serrated
	Splenius	Bandage shaped
	Teres	Long and round
	Trapezius	Trapezoidal
Location or direction relative to body axes	Anterior	Front
	Externus	Superficial
	Extrinsic	Outside
	Inferioris	Inferior
	Internus	Internal or deep
	Intrinsic	Inside
	Lateralis	Lateral
	Medialis	Medial or middle
	Medius	Medial or middle
	Obliquus	Oblique

(continued)

Table 5.1   **Muscle Terminology** (continued)

Factor	Term	Meaning
Location or direction relative to body axes (continued)	Posterior	Back
	Profundus	Deep
	Rectus	Straight or parallel
	Superficialis	Superficial
	Superioris	Superior
	Transversus	Transverse
Location relative to body region	Abdominis	Abdomen
	Anconeus	Elbow
	Brachialis	Brachium (upper arm)
	Capitis	Head
	Capri	Wrist
	Cervicis	Neck
	Cleido/clavius	Clavicle
	Costalis	Ribs
	Cutaneous	Skin
	Femoris	Femur
	Glosso/glossal	Tongue
	Hallucis/hallux	Big (great) toe
	Ilio-	Ilium
	Lumborum	Lumbar region
	Nasalis	Nose
	Oculo-	Eye
	Oris	Mouth
	Pollicis	Thumb
	Popliteus	Behind the knee
	Radialis	Radius
	Scapularis	Scapula
	Temporalis	Temples
	Thoracis	Thoracic region
	Tibialis	Tibia
	Ulnaris	Ulna
Action	Abductor	
	Adductor	
	Depressor	
	Extensor	
	Flexor	
	Levator	
	Pronator	
	Rotator	
	Supinator	
	Tensor	
Number of origins or muscle heads/bellies	Biceps	Two heads
	Triceps	Three heads
	Quadriceps	Four heads

Muscles sometimes are collectively referred to by group names. When using group names, keep in mind that the group name represents a collection of muscles rather than a single muscle. For example, quadriceps, or quads, is a group name for four muscles on the anterior thigh. The quadriceps group includes the vastus medialis, vastus lateralis, vastus intermedius, and rectus femoris. Similarly, the hamstring group contains the biceps femoris, semitendinosus, and semimembranosus. Other common group names include the rotator cuff group (subscapularis, supraspinatus, infraspinatus, and teres minor) and the triceps surae group (soleus and

In Others' Words: Kenneth D. Keele

Kenneth D. Keele, in his definitive treatise on the scientific contributions of Leonardo da Vinci, concludes that "of all the 'powers' of nature, movement was to Leonardo the most basic and the most fascinating. And of all the powers of the human body, movement, with its changes of shape or 'mutations,' was at the very heart of his lifelong activities. Studies of the muscles bringing about geometrical mutations of the human body occupied a great deal of his time. It is no exaggeration to say that some hundreds of pages of his notes are devoted to some aspect of muscular action. Leonardo was, one may say, dissecting the movements of the human body rather than a motionless dead cadaver" (Keele, 1983, p. 267).

—Kenneth D. Keele, Leonardo da Vinci's Elements of the Science of Man

gastrocnemius). Careful consideration of the last group name, the triceps surae, may seem incongruous with the fact that the group contains only two muscles while the name (triceps) would suggest three. This is reconciled by noting that the gastrocnemius has two heads (medial and lateral). The two heads of the gastrocnemius together with the single head of the soleus form a "triceps" group, or triceps surae.

When learning the many muscles of the human body, attention paid to clues within a muscle's name can facilitate the learning process. In addition, try to *visualize* each muscle's location, attachments, and movements as you study. Being able to see the muscle and its action can be helpful in learning about the muscular system and its function. Students who only memorize the tables and lists place themselves at a disadvantage and make the task of learning anatomy more difficult.

Functional Actions of Muscles

You can determine the action of a muscle in several ways. One of the simplest methods, termed **palpation**, involves feeling the muscle action through the skin. When you flex your elbow, for example, the bulging of the biceps brachii is easily palpated. Obviously, palpation works only for superficial muscles, not for deep muscles that are obscured by different muscles or other tissues.

Another method involves assessment of a muscle's location and its attachments. The action of a muscle can be deduced by constructing a simple line drawing of the muscle and its **origin** and **insertion** (figure 5.2). Draw a straight line, or if necessary a curved line (e.g., gluteus

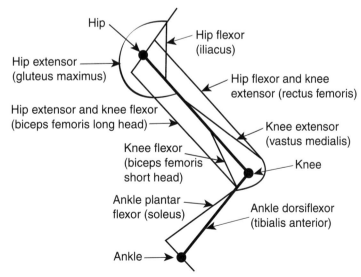

Figure 5.2 Model of one-joint and two-joint muscles.

Adapted from I. Schenau, Bobbert, and Soest, 1990.

maximus), connecting the origin to the insertion. Recall that the origin usually remains relatively fixed (immovable) and that the insertion typically moves toward the origin. Place an arrow along the muscle line of action you have drawn, with the head of the arrow directed toward the origin. Move the bone attached at the insertion toward the bone attached to the origin, and observe the movement produced. It's just that simple.

For biarticular (two-joint) muscles (e.g., semitendinosus, rectus femoris) repeat the procedure, but this time direct the arrow toward the insertion. Now rotate the bone with the origin attachment toward the bone of insertion attachment, and observe the motion at the second joint .

When you can visualize joint movements in this way, you are far along the road to understanding human movement. Visualization of movement is much easier than rote memorization. Learning muscles and their functions will be easier if you say the names, write down their attachments and functions, and if possible, palpate the muscles on your own body. Most important, do your best to create vivid images in your mind and "see" the movements through your own imagination.

Another way of assessing muscle function is through the use of **electromyography** (EMG). EMG measures the electrical activity of muscles and allows researchers and clinicians to explore the action of muscles in a wide variety of movements. Much of the information on muscle action presented in this and later chapters comes from electromyography studies.

Before we continue, commit to memory the following notes on terminology and presentation format:

- Muscles are said to "cross" a joint, meaning they have action, or can produce movement, at the joint being crossed. Some muscles have action at a single joint and are termed **uniarticular**. Other muscles cross two joints (**biarticular**), three joints (**triarticular**), or more than three joints (**multiarticular**). Biarticular muscles are common in the human body, and their function is explored in greater detail in chapter 7. Triarticular muscles are rare. Multiarticular muscles are primarily found in the distal regions of the limbs (e.g., muscles that move the fingers and toes).

- Joint actions, or movements, typically are described in one of two ways: by identifying either the joint being moved (e.g., flexion of the elbow) or the segment being moved (e.g., flexion of the forearm). Both methods are correct, and each is used by authors of other texts.

- Muscle actions refer to what each muscle does when acting concentrically.

- All muscles acting at a particular joint do not participate equally in performing or controlling a movement. The relative contribution of each muscle is determined by many factors, including muscle size (cross-sectional area), level of neural stimulation, muscle fiber type composition, fatigue, joint position, and the type of movement being performed. The muscles making the greatest contribution are termed **prime movers**. In plantar flexion at the ankle, for example, the relatively large gastrocnemius and soleus play a much greater role than does the diminutive plantaris.

Muscles of Major Joints

To see how muscles act together, we now present in tables 5.2 through 5.6 and figures 5.3 through 5.7 the primary movements at each of the major joints in the body and the muscles acting to produce and control those movements. (Note that in the tables, nonitalicized muscles are considered prime movers. Italicized muscles represent assistant movers.) As a reminder, the muscle actions described are for *concentric* action. The *eccentric* action of a muscle controls the opposite motion listed for its respective concentric action. For example, the biceps brachii acts concentrically to produce elbow flexion and works eccentrically to control elbow extension.

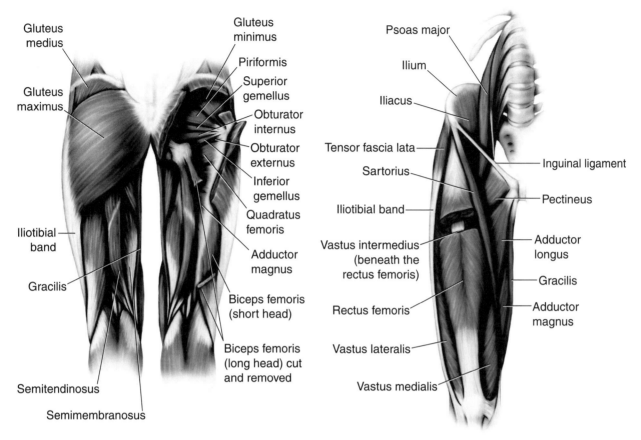

Figure 5.3 Muscles of the hip and knee joints.

Table 5.2 Hip and Knee Joint Muscle Actions

Hip joint	
Extension	**Flexion**
Gluteus maximus	Psoas major
Semitendinosus	Iliacus
Semimembranosus	Pectineus
Biceps femoris, long head	Rectus femoris
Adductor magnus, posterior fibers	Adductor brevis
	Adductor longus
	Adductor magnus, anterior upper fibers
	Tensor fascia lata
	Sartorius
	Gracilis
Abduction	**Adduction**
Gluteus medius	Pectineus
Gluteus minimus	Adductor brevis
Tensor fascia lata	Adductor longus
Gluteus maximus, superior fibers	Adductor magnus
Psoas major	Gracilis
Iliacus	*Gluteus maximus, inferior fibers*
Sartorius	

(continued)

Table 5.2 **Hip and Knee Joint Muscle Actions** (continued)

Hip joint	
Medial rotation	Lateral rotation
Gluteus minimus	Gluteus maximus
Tensor fascia lata	Piriformis
Pectineus	Gemellus superior
Adductor brevis	Obturator internus
Adductor longus	Gemellus inferior
Adductor magnus, anterior upper fibers	Obturator externus
Semitendinosus	Quadratus femoris
Semimembranosus	*Psoas major*
	Iliacus
	Sartorius
	Biceps femoris, long head

Knee joint	
Extension	Flexion
Vastus medialis	Semimembranosus
Vastus intermedius	Semitendinosus
Vastus lateralis	Biceps femoris
Rectus femoris	*Sartorius*
	Gracilis
	Popliteus
	Gastrocnemius
	Plantaris

Medial rotation[1]	Lateral rotation[1]
Popliteus	Biceps femoris
Semimembranosus	
Semitendinosus	
Sartorius	
Gracilis	

[1]Rotation can occur only when the knee is flexed.

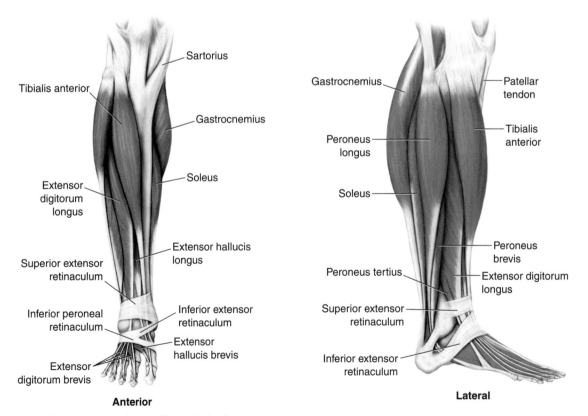

Anterior

Lateral

Figure 5.4 Muscles of the ankle and subtalar joints.

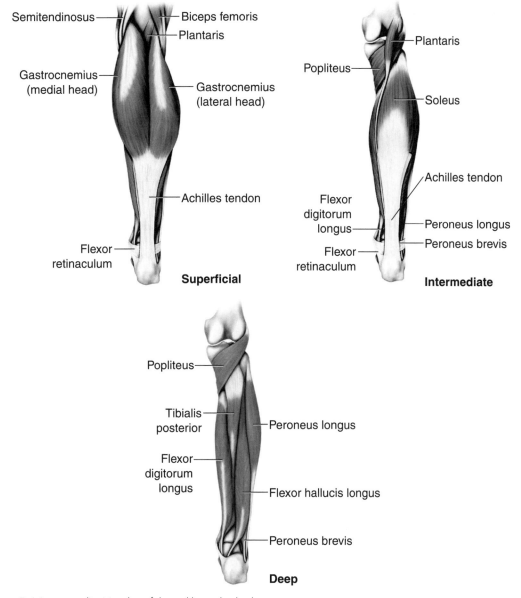

Semitendinosus
Biceps femoris
Plantaris
Gastrocnemius (medial head)
Gastrocnemius (lateral head)
Achilles tendon
Flexor retinaculum

Superficial

Plantaris
Popliteus
Soleus
Achilles tendon
Flexor digitorum longus
Peroneus longus
Peroneus brevis
Flexor retinaculum

Intermediate

Popliteus
Tibialis posterior
Peroneus longus
Flexor digitorum longus
Flexor hallucis longus
Peroneus brevis

Deep

Figure 5.4 *(continued)* Muscles of the ankle and subtalar joints.

Table 5.3 Ankle and Subtalar Joint Muscle Actions

Ankle joint	
Plantar flexion	Dorsiflexion
Gastrocnemius	Tibialis anterior
Soleus	Extensor digitorum longus
Plantaris	Peroneus tertius
Tibialis posterior	*Extensor hallucis longus*
Flexor hallucis longus	
Flexor digitorum longus	
Peroneus longus	
Peroneus brevis	

(continued)

Table 5.3 Ankle and Subtalar Joint Muscle Actions (continued)

Subtalar joint	
Inversion	Eversion
Tibialis anterior	Peroneus longus
Tibialis posterior	Peroneus brevis
Flexor hallucis longus	Peroneus tertius
Flexor digitorum longus	Extensor digitorum longus
Gastrocnemius	
Soleus	
Plantaris	

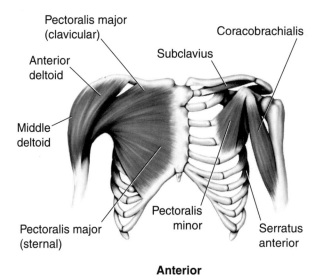

Figure 5.5 Muscles of the shoulder girdle and shoulder joint.

Table 5.4 Shoulder Girdle and Shoulder Joint Muscle Actions

Shoulder girdle	
Elevation	Depression
Levator scapula	Lower trapezius
Upper trapezius	Pectoralis minor
Rhomboids	
Adduction	**Abduction**
Rhomboids	Pectoralis minor
Middle trapezius	Serratus anterior
Upward rotation	**Downward rotation**
Trapezius	Rhomboids
Serratus anterior	Levator scapula
	Pectoralis minor
Shoulder (glenohumeral) joint	
Flexion	Extension
Pectoralis major, clavicular portion	Pectoralis major, sternal portion
Anterior deltoid	Latissimus dorsi
Biceps brachii, short head	Teres major

Shoulder (glenohumeral) joint *(continued)*	
Flexion	**Extension**
Coracobrachialis	*Posterior deltoid*
	Triceps brachii, long head
Adduction	**Abduction**
Latissimus dorsi	Middle deltoid
Teres major	Supraspinatus
Pectoralis major, sternal portion	Anterior deltoid
Biceps brachii, short head	*Biceps brachii*
Triceps brachii, long head	
Medial (internal) rotation	**Lateral (external) rotation**
Latissimus dorsi	Teres minor
Teres major	Infraspinatus
Subscapularis	*Posterior deltoid*
Anterior deltoid	
Pectoralis major	
Biceps brachii, short head	
Horizontal flexion (adduction)	**Horizontal extension (abduction)**
Pectoralis major	Middle deltoid
Anterior deltoid	Posterior deltoid
Coracobrachialis	Teres minor
Biceps brachii, short head	Infraspinatus
	Teres major
	Latissimus dorsi

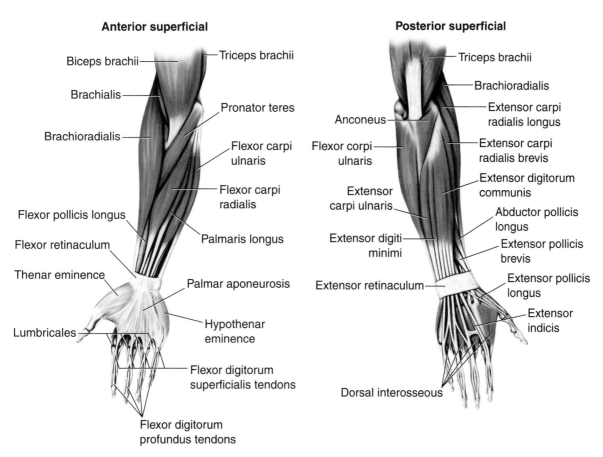

Anterior superficial

Biceps brachii
Brachialis
Brachioradialis
Triceps brachii
Pronator teres
Flexor carpi ulnaris
Flexor carpi radialis
Flexor pollicis longus
Flexor retinaculum
Thenar eminence
Palmaris longus
Palmar aponeurosis
Lumbricales
Hypothenar eminence
Flexor digitorum superficialis tendons
Flexor digitorum profundus tendons

Posterior superficial

Triceps brachii
Brachioradialis
Extensor carpi radialis longus
Anconeus
Flexor corpi ulnaris
Extensor carpi radialis brevis
Extensor carpi ulnaris
Extensor digitorum communis
Extensor digiti minimi
Abductor pollicis longus
Extensor pollicis brevis
Extensor retinaculum
Extensor pollicis longus
Extensor indicis
Dorsal interosseous

Figure 5.6 Muscles of the elbow, radioulnar, and wrist joints.

Table 5.5   Elbow, Radioulnar, and Wrist Joint Muscle Actions

Elbow joint	
Flexion	**Extension**
Biceps brachii	Triceps brachii
Brachialis	*Anconeus*
Brachioradialis	

Radioulnar joint	
Supination	**Pronation**
Biceps brachii	Pronator teres
Supinator	Pronator quadratus
Brachioradialis*	*Brachioradialis**
*functions to move the forearm to the mid- or neutral position	*functions to move the forearm to the mid- or neutral position

Wrist joint	
Flexion	**Extension**
Flexor carpi radialis	Extensor carpi radialis longus
Flexor carpi ulnaris	Extensor carpi radialis brevis
Flexor digitorum superficialis	Extensor carpi ulnaris
Flexor digitorum profundus	*Extensor indicis*
Palmaris longus	*Extensor digiti minimi*
	Extensor digitorum

Radial deviation (abduction)	**Ulnar deviation (adduction)**
Flexor carpi radialis	Flexor carpi ulnaris
Extensor carpi radialis longus	Extensor carpi ulnaris
Extensor carpi radialis brevis	

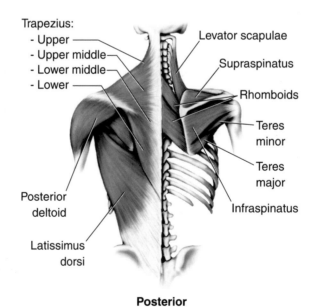

Posterior

Figure 5.7   Muscles of the vertebral column.

Table 5.6   Vertebral Column Muscle Actions

Vertebral column (thoracic and lumbar regions)	
Flexion	**Extension**
Rectus abdominis External oblique Internal oblique *Psoas major (lumbar region)*	Erector spinae group
Rotation to the same side	**Rotation to the opposite side**
Internal oblique	External oblique
Lateral flexion	
External oblique Internal oblique Quadratus lumborum *Rectus abdominis*	

concluding comments

Our gross movements are controlled by a relatively small subset of the body's more than 600 skeletal muscles. Sets of muscles at each major joint in the body work together in coordinated fashion to produce and control our movements. In most cases, they do so with little conscious attention on our part. Our neuromuscular system's ability to control movements is remarkable. Often it is only when something affects our ability to move, such as injury, and our regular movements are disrupted that we become aware of how effectively our muscles work on a day-to-day basis.

critical thinking questions

1. Muscle names often describe their size, shape, location, action, or attachment sites, as well as the number of muscle heads (bellies) or the direction of their fiber orientation. List specific muscles whose names contain one or more of these descriptive terms.

2. Describe the diagrammatic technique presented in this chapter for determining a muscle's actions. Select at least two muscles that function across more than one joint and apply the technique. Include a drawing to support your answers.

3. Describe some of the principal factors that affect a muscle's relative contribution to a specific movement. In addition, when analyzing a group of muscles performing a specific action, what distinguishes prime from assistant movers?

4. Muscles' actions are defined by their concentric actions. What do we mean when we say a muscle's eccentric action controls, rather than produces, the opposite action (motion)?

suggested readings

Behnke, R.S. (2001). *Kinetic anatomy*. Champaign, IL: Human Kinetics.

Estes, S.G., & Mechikoff, R.A. (1999). *Knowing human movement*. Boston: Allyn & Bacon.

Jenkins, D.B. (1998). *Hollinshead's functional anatomy of the limbs and back* (7th ed.). Philadelphia: Saunders.

Levangie, P.K., & Norkin, C.C. (2001). *Joint structure and function: A comprehensive analysis* (3rd ed.). Philadelphia: Davis.

MacKinnon, P., & Morris, J. (1994). *Oxford textbook of functional anatomy* (Vol. 1, rev. ed.). Oxford, UK: Oxford University Press.

Martini, F.H., Timmons, R.J., & McKinley, M.P. (2000). *Human anatomy* (3rd ed.). Upper Saddle River, NJ: Prentice Hall.

Platzer, W. (1992). *Locomotor system*. New York: Thieme Medical.

Simons, D.G., Travell, J.G., Simons, L.S., & Cummings, B.D. (1998). *Travell & Simons' myofascial pain and dysfunction: The trigger point manual* (2nd ed.). Philadelphia: Lippincott Williams & Wilkins.

Thompson, C.W., & Floyd, R.T. (1994). *Manual of structural kinesiology* (12th ed.). St. Louis: Mosby.

Travell, J.G., Simons, D.G., & Cummings, B.D. (1992). *Myofascial pain and dysfunction: The trigger point manual, vol. 2: The lower extremities*. Philadelphia: Lippincott Williams & Wilkins.

Warfel, J.H. (1993). *The extremities: Muscles and motor points* (6th ed.). Philadelphia: Lea & Febiger.

Warfel, J.H. (1993). *The head, neck, and trunk* (6th ed.). Philadelphia: Lea & Febiger.

Applied Dynatomy

Human movement must obey the laws of mechanics as set forth by Sir Isaac Newton and others. We therefore begin part II with chapter 6 on the mechanics of movement that presents a qualitative summary of essential mechanical terms and concepts. With that foundation in place, we proceed in chapter 7 to a discussion of muscular control of movement and movement assessment. This chapter presents the *muscle control formula*, a set of simple steps that identifies the muscles responsible for producing and controlling any human movement. In addition, we consider movement concepts and assessment techniques important to a full understanding of human movement.

The next three chapters examine the dynatomy of everyday movements. Chapter 8 presents the basics of posture and balance, and walking. Chapter 9 continues with discussion of running, jumping, throwing, kicking, and lifting. We then apply the dynatomy approach in chapter 10 to exemplar movements in resistance training, cycling, swimming, and dance. We conclude with a brief chapter in which we outline areas of importance in the future of human movement studies.

six

Mechanics of Movement

objectives

After studying this chapter, you will be able to do the following:

- Identify the major areas of biomechanics relevant to human movement: movement mechanics, fluid mechanics, joint mechanics, and material mechanics

- Explain general biomechanical concepts and measures, including linear and angular motion, center of gravity, stability, mobility, and movement equilibrium

- Explain concepts of movement mechanics, including kinematics, kinetics, force, pressure, lever systems, torque (moment of force), Newton's laws of motion, work, power, energy, momentum, and friction

- Explain concepts of fluid mechanics, including fluid flow and resistance

- Explain concepts of joint mechanics, including range of motion, joint stability, and joint mobility

- Explain concepts of material mechanics, including stress, strain, stiffness, bending, torsion, and viscoelasticity

Leonardo da Vinci "himself endorses . . . that by means of studying his Elements of Mechanics 'you will be able to prove all your [anatomical] propositions.'"

Kenneth D. Keele, Leonardo da Vinci's Elements of the Science of Man

Movement fascinates us today in much the same way as it has our ancestors over many millennia. Artists have used human movement as inspiration for artistic expression. Scientists have made human movement the subject of extensive scientific inquiry. From a scientific point of view, movement can be considered from many perspectives, including those of anatomy, physiology, psychology, and physics. One branch of physics, mechanics, is particularly applicable to the assessment and appreciation of movement. With respect to human movement, the mechanical perspective falls within the domain of **biomechanics**, broadly defined as the application of mechanical principles to the study of biological organisms and systems.

Why is the study of biomechanics important in understanding human movement? Perhaps the most important reason is that our movement potential and limitations often are dictated by mechanical properties and events. How fast we can run, how high we can jump, and how much we can lift are determined by the forces acting both inside and outside our bodies. Many of the details of how biomechanics affects human movement are presented in this chapter, with a continuing theme of mechanical influence evident throughout the following chapters as well.

This chapter considers several areas of biomechanics that are relevant to understanding the dynamics of human movement. These areas include some basic biomechanical concepts, followed by brief discussion of movement mechanics, fluid mechanics, joint mechanics, and material mechanics.

Biomechanical Concepts

Many excellent texts are devoted solely to the discussion of human biomechanics. In a single chapter, we are limited to presenting only a brief summary of mechanical concepts most relevant to understanding human movement. Although biomechanics is inherently a quantitative discipline, the approach taken here is to limit mathematical considerations and focus instead on providing a conceptual framework for understanding the mechanics of movement.

In discussing biomechanics, we often refer to objects as bodies. The word **body** in this context is taken to mean any collection of matter. It may refer to the entire human body, a segment (e.g., thigh or upper arm), or any other collected mass (e.g., a block of wood).

The amount of matter, or substance, constituting a body is known as its **mass**. A body's mass is *not* the same as its **weight**. Mass is certainly related to weight, but it is important to understand that the two terms, *mass* and *weight*, should not be used as synonyms. The relationship between mass and weight is shown in figure 6.1.

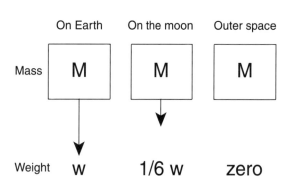

Figure 6.1   Relationship between mass and weight.

Linear and Angular Motion

In mechanical terms there are two basic forms of movement: (1) **linear motion**, in which a body moves along a straight line (**rectilinear motion**) or a curved line (**curvilinear motion**), and (2) **angular motion**, or **rotational motion**, in which a body rotates about a fixed line called the axis of rotation. In walking, for example, the thigh rotates in the sagittal plane about an axis of rotation through the hip, the lower leg (shank) rotates about an axis through the knee, and the foot rotates about an axis through the ankle (figure 6.2).

Many human movements combine linear and angular motion in what is termed **general motion**. The movement of a person's thigh during walking, for example, involves linear motion of the entire thigh segment in a forward direction and angular motion at the hip in alternating periods of flexion and extension.

The movement of inanimate objects can also exhibit general motion. The flight of a thrown softball, for example, consists of linear motion (the curved path, or arc, of the ball) and angular motion (the ball's spin).

As we explore the mechanics of movement, the concepts of linear and angular motion recur often. Combination of these two simple movement forms results in the vast array of movement patterns we perform and observe on a daily basis.

Center of Gravity

Every body contains a point, known as the **center of mass** or **center of gravity**, about which that body's mass is equally distributed. Although

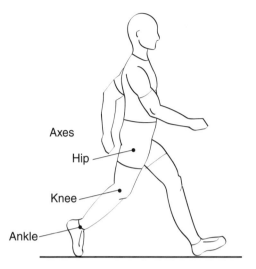

Figure 6.2 Lower-extremity joint axes at the hip, knee, and ankle during walking.

there is a technical difference in the definitions of center of mass and center of gravity, in practical terms the two points are coincident, and we therefore use the terms interchangeably.

If the body's mass were concentrated into a **point mass** at the center of gravity, this point would move in exactly the same way as the body would in its original distributed state. This concept is important in discussing human movement because all points within a given body often are not moving in the same direction or at the same speed. Figure 6.3, for example, shows a gymnast performing a vault. The twists and turns she executes happen around her center of gravity while her center of gravity moves along a smooth curve, or arc.

In normal standing, the body's center of gravity is located just anterior to the second sacral vertebra. From a frontal view, the center of gravity normally is located along the body's midline and is typically 55 to 57% of a person's height above the ground (figure 6.4).

Movement of body segments (e.g., lifting the arms, flexing the knee) causes the center of gravity to move within the body. For example, raising (abducting) the arms from anatomical position to an overhead position results in superior movement of the center of gravity along the body's midline. If only one arm is raised, the center of gravity similarly moves superiorly but also deviates from the midline slightly toward the raised-arm side.

Most of the time, the center of gravity lies within the body. However, in the same way as a doughnut's center of gravity lies outside the "body" of the doughnut (i.e., in the "hole"),

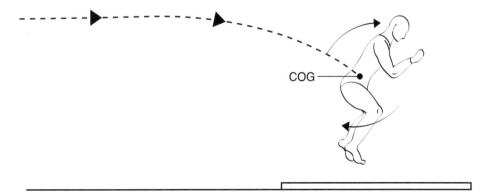

Figure 6.3 Linear and angular motion as seen in a gymnast during her dismount.

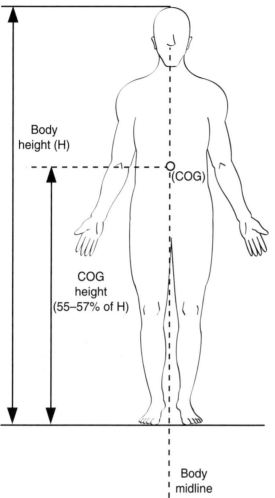

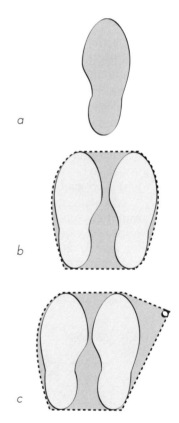

Figure 6.5 Center of gravity located outside the body when the body is in a bent-over, or pike, position.

Figure 6.4 Center of gravity location in the human body.

the center of gravity of a human body can lie outside the body when a person assumes a bent-over, or pike, position, as in gymnastics or diving (figure 6.5).

Stability, Mobility, and Movement Equilibrium

When standing, we typically have two feet in contact with the ground. If our feet are close together, we feel less stable than when the feet are spread apart. Increasing the distance between the feet increases what is termed our **base of support**, defined as the area within an outline of all ground contact points. Several bases of support are illustrated in figure 6.6. The importance of the base of support in determining our stability and ability to move effectively is discussed in this section and again, in more detail, in chapters 8 and 9, where we explore movements such as walking, running, jumping, and throwing.

As we prepare to move in any situation, with little or no conscious thought we place ourselves along what might be termed a *stability–mobility continuum*. In situations of imminent contact,

Figure 6.6 Examples of bases of support as seen from above: (a) one-legged standing, (b) two-legged standing, and (c) standing with a cane.

we try to enhance our **stability**; when we want to move quickly, we try to increase our **mobility**. In preparation for impending contact by an opponent, for example, an American football player will try to brace himself by widening his base of support and bending his knees. If, on the other hand, the player decides to run away from the collision, he would adopt a different body posture that would enhance his mobility.

From a mechanical perspective, five factors determine our levels of stability and mobility.

- *Size of the base of support in the direction of force or impending force:* In general, increasing the size of the base of support increases stability. In preparation for an impact, we tend to spread our feet apart. We do so, however, in the direction of the force. If you were about to be struck from the front, would you widen your base of support by abducting at the hips to spread your feet apart to the side? Probably not. Most likely you would increase your base of support by staggering your feet front-to-back, or in an anterior–posterior orientation. Merely increasing the size of your base of support will not necessarily make you more stable. The increase must be made in the direction of force or impending force. Increases in base of support can be made by placing the feet in a certain position, as in the previous example, or by adding ground contact points. Additional contact points can be added by using other body parts, as when a baby creeps along the ground on hands and knees or when an athlete assumes a three-point or four-point stance (figure 6.7). Older or injured persons also can enhance their stability by using a cane or crutch to add contact points to the system, thereby increasing their bases of support (see figure 6.6c).

- *Height of the center of gravity above the base of support:* When you squat down to improve your stability, you lower your center of gravity, or decrease the height of the center of gravity above the base of support. Conversely, standing up straight raises the center of gravity above the base of support and decreases stability.

- *Location of the center of gravity projection within the base of support:* Imagine that you drop a plumb line (i.e., a string with a weight on the end) straight down from your center of gravity. That line is referred to as the *vertical projection*, or **projection**, of your center of gravity within the base of support. If the projection moves outside the base of support, you become very unstable and will fall without corrective muscle action. In normal standing, when the center of gravity projection lies at or near the center of the base of support, you are more stable than when the projection lies near the edge of the base of support. When another body is about to collide with yours, you tend to lean toward the colliding body. This lean moves the projection near the edge of the base of support so that at impact, the center of gravity has a greater distance to travel before leaving the base of support on the opposite side and causing you to fall.

- *Body mass or body weight:* A body's mass (or weight) contributes to stability. Simply stated, heavier bodies are harder to move and hence are more stable. Lighter bodies are moved more easily and are less stable.

- *Friction:* The amount of frictional resistance at the interface between the ground and any contact points (e.g., foot or shoe) contributes to stability and mobility. A young basketball player

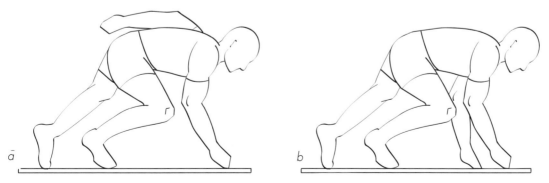

Figure 6.7   Examples of *(a)* three-point and *(b)* four-point stances.

trying out her new shoes on a freshly polished gymnasium floor would encounter relatively high friction that would improve her stability. A teenager running on an icy sidewalk in the middle of winter, in contrast, would have much lower stability because of the low friction and would be more likely to slip and fall.

In summary, high stability (low mobility) is characterized by a large base of support, a low center of gravity, a centralized center of gravity projection within the base of support, a large body mass, and high friction at the ground interface. Low stability (high mobility), in contrast, occurs with a small base of support, a high center of gravity, a center of gravity projection near the edge of the base of support, a small body mass, and low friction.

Movement Mechanics

Human movement results from mechanical factors that produce and control movement from inside the body *(internal mechanics)* or affect the body from without *(external mechanics)*. Examples of internal mechanical factors include the forces produced by muscle action and the joint stability provided by ligaments. External mechanical factors include gravity, air resistance, and other external forces acting on the body.

Movement can be assessed descriptively or by investigation of the underlying forces. The description of spatial and timing characteristics of movement without regard to the forces involved is known as **kinematics**. The assessment of movement with respect to the forces involved is called **kinetics**.

Kinematics

Kinematics involves the description of movement with respect to space and time without regard to the forces that produce or control the movement. Kinematics involves five primary variables:

- Timing, or temporal, characteristics of movement
- Position or location
- Displacement (measuring the movement from starting point to ending point)
- Velocity (a measure of how fast something moves)
- Acceleration (an indicator of how quickly the velocity changes)

Kinematics can be considered for movements viewed two dimensionally or three dimensionally. Some movement patterns, such as walking, are essentially planar, so two-dimensional assessment is sufficient. Other patterns, such as throwing, are multiplanar movements requiring three-dimensional consideration.

- *Time:* The kinematic variable of time provides a measure of the duration of a particular event. Noting that during a single step a person's foot is in contact with the ground for 0.5 seconds would be an example of such a timing (temporal) measure. The duration, or time, may be long (e.g., a marathon that may take several hours) or very short (e.g., the duration of force application when a soccer player kicks the ball).

- *Position:* The position of a person's whole body, or a segment of the body, at any instant in time plays an important role in determining the mechanical characteristics of the body system. The position of a body segment can be described qualitatively (e.g., leg is abducted) or quantitatively (e.g., forearm is positioned with the elbow flexed 90°).

- *Displacement:* When a body moves from one place to another, we measure this displacement in a straight line from the starting point to the ending position. This is termed the **linear displacement**. A body rotating about an axis experiences **angular displacement**, which is measured as the number of degrees of rotation (e.g., the knee flexed through an angular displacement of 35°). In kinematics, a distinction is drawn between displacement and distance.

As already defined, displacement is measured along a straight line from one point to another. **Distance**, in contrast, represents the overall measure of how far the body moves in getting from a starting point to a finishing point. Figure 6.8 shows the difference between displacement and distance.

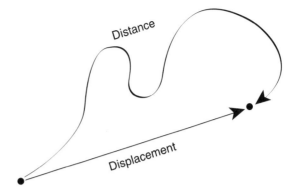

Figure 6.8   Displacement versus distance.

• *Velocity:* Velocity is a measure of how fast the body is moving and in what direction. For example, an Olympic sprinter might run with a linear velocity of 10 m/s in a straight line from the starting blocks to the finish line. Velocity can also be used to measure angular motion, as when a softball pitcher swings her arm with an angular velocity of 1,000 degrees per second in a counterclockwise direction. In common usage, the terms *velocity* and *speed* often are used interchangeably. In mechanical terms, however, they have distinct—though related—meanings. Velocity is a vector quantity (magnitude and direction), while speed is a scalar (magnitude only) measure. The speed of a runner might be 5 m/s. To transform the movement measurement to velocity, we must indicate the running direction (e.g., 5 m/s due north).

• *Acceleration:* Acceleration measures the change in a body's velocity. Linear acceleration is measured as the change in linear velocity divided by the change in time. Similarly, angular acceleration is the change in angular velocity divided by the change in time. One of the most common accelerations affecting human movement is the acceleration due to gravity. Gravitational acceleration acts downward, as seen in the simple example of dropping a ball. When the ball is being held, its velocity is zero. Once dropped, the ball accelerates by increasing its downward velocity. As long as gravity is acting, the ball falls faster and faster until it reaches terminal velocity. Linear acceleration often is expressed in units of *g*s, where 1 *g* is the acceleration created by the earth's gravitational pull (~9.81 m/s^2). Thus, if a boxer's head, struck by his opponent's punch, experiences 5 *g*s, the boxer has been hit by a force strong enough to accelerate the head at five times the acceleration caused by the force of gravity.

Kinetics

Kinematic description is an important first step in analyzing any movement. Kinematic analyses, however, are limited to describing the spatial and timing characteristics of movement without investigating the underlying forces involved. Because forces cause movement, kinetics (the study of forces and their effects) is an area worthy of consideration. We now present important force-related concepts.

Inertia and Moment of Inertia

Bodies at rest have a tendency to resist being moved. Because there are two forms of movement (linear and angular), there logically are two forms of resistance to motion. The resistance to linear motion is termed inertia, while the resistance to angular motion is termed moment of inertia.

As noted earlier, mass is the quantity of matter. In SI units (système international d'unités), it is measured in kilograms (kg). Common sense suggests that the greater a body's mass, the more difficult it is to move. This resistance to being moved linearly, termed **inertia**, is defined as the property of matter by which it remains at rest or in motion. To linearly move an object at rest, we must overcome its inertia, or its tendency to stay where it is. To slide a box from a resting position across the floor, you must push or pull it with enough force to overcome its inertia.

In the same way that bodies resist linear motion, they also tend to resist all forms of angular movement, including rotation, bending, and twisting (torsion). The general term used to describe this resistance to a change in angular state of position or motion is **moment of inertia**. A word of caution is warranted here because confusion sometimes arises from using the word *inertia* (which we just defined as resistance to change in a body's state of linear position or motion) in the term *moment of inertia*, used to describe angular resistance.

Three types of moment of inertia correspond with three forms of angular movement. The first, **mass moment of inertia**, describes the resistance to a body's rotation about an axis. A body at rest with a fixed axis (e.g., a pendulum) will resist being moved rotationally, just as a body that is already rotating at a constant angular velocity will tend to maintain that angular velocity and will resist change in its velocity. This concept holds, of course, for human movements. In normal standing, for example, the arms hang straight down. To abduct or flex the limbs from this position, their resistance to being moved angularly (i.e., their mass moment of inertia) must be overcome by muscle action at the shoulder (glenohumeral) joints.

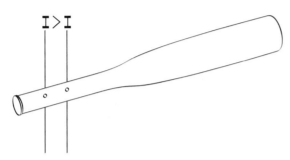

Recall that the magnitude of resistance in the case of linear movement is determined by the mass of the object. In the case of angular movement, the magnitude of the resistance is determined by the mass and the mass's location, or distribution, with respect to the axis of rotation. As the mass is moved farther from the axis, the resistance, or mass moment of inertia, increases (figure 6.9).

Figure 6.9 Effect of choking up on a bat on the mass moment of inertia.

Moving a limb segment's mass closer to or farther away from a joint axis has a profound effect on overall movement. A sprinter, for example, when swinging his leg through the air in preparation for the foot hitting the ground, will flex his knee as much as possible to bring the combined mass of the lower leg and foot closer to the hip joint axis of rotation, thus reducing the resistance to rotation (mass moment of inertia) and allowing the entire leg to swing through its arc as quickly as possible.

The second form of angular resistance, **area moment of inertia**, is the resistance to bending. Because our limb segments are relatively rigid (i.e., they bend very little) during most normal movements, the area moment of inertia is of minor concern. In case of injury, however, when a segment and its enclosed bones are severely bent, the area moment of inertia plays a prominent role. If the bones have insufficient strength to resist the bending, they will fracture.

The third form of resistance, **polar moment of inertia**, describes the resistance to twisting, or torsion. As with bending, torsional loading of segments is usually quite low, and the effects of the polar moment of inertia are minimal. But consider the case of a skier whose ski is violently twisted during a fall. The twisting ski transfers a torsional force to the lower leg and may result in fracture of the tibia. In this case, the polar moment of inertia of the tibia determines whether injury occurs and how severe the fracture will be.

Subsequent sections on material mechanics discuss the resistance to bending and torsion more fully.

Force

Force, a fundamental element in human movement mechanics, is defined as a mechanical action or effect applied to a body that tends to produce acceleration. In simpler terms, force can be described as a push or pull. The standard (SI) unit of force is the newton (N), defined as the force required to accelerate a 1 kg mass at 1 m/s² in the direction of the force (1 N = 1 kg · m · s⁻²). One pound of force equals 4.45 N, so someone who weighs 180 lb as measured in British units would weigh 801 N in SI units.

As a prelude to a more general discussion of force, we introduce the concept of an **idealized force vector**. If we consider, for example, the forces acting on the head of the femur while a person assumes a standing posture, an infinite number of individual force vectors could be distributed over the joint surface. Each of these vectors would have its own magnitude and direction. To analyze the effect of all these vectors would be a time-consuming task, to say the least, requiring sophisticated instrumentation and computer modeling to accomplish. We can create a single force vector (idealized force vector) that represents the net effect of all the other vectors, essentially idealizing the situation through simplification. What is lost in information describing the distribution of forces is gained by creating a model with a single vector from which calculations and evaluations can be made.

This notion of an idealized force vector is useful in many situations. For example, consider the concept of a body's center of gravity. Using an idealized force vector, we can represent the weight of the body by a single vector projecting down from the body's center of gravity (figure 6.10).

The forces inherent to movement analysis are those that act in or on the human body. Among these are the force of gravity (a downward force tending to accelerate objects at ~9.81 m/s²), the impact of the feet on the ground, objects colliding with the body (e.g., a thrown ball or another body), musculotendinous forces, frictional forces with the ground, ligament forces acting to stabilize joints, and compressive forces exerted on long bones of the lower extremities.

The result of forces on movement depends on the combined effect of seven force-related factors (Whiting & Zernicke, 1998):

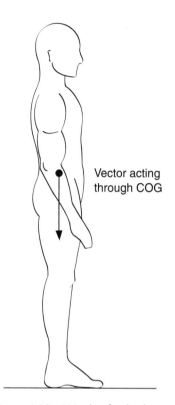

Vector acting through COG

Figure 6.10 Weight of a body represented by a single idealized force vector acting at the body's center of gravity.

Reprinted, by permission, from W.C. Whiting and R.F. Zernicke, 1998, *Biomechanics of musculoskeletal injury* (Champaign, IL: Human Kinetics), 48.

- Magnitude (How much force is applied?)
- Location (Where on the body or structure is the force applied?)
- Direction (Where is the force directed?)
- Duration (Over what time interval is the force applied?)
- Frequency (How often is the force applied?)
- Variability (Is the magnitude of the force constant or variable over the application interval?)
- Rate (How quickly is the force applied?)

Pressure

In many situations, it is important to know how the force is distributed over the surface being contacted. In walking, for example, the area of foot contact (e.g., heel versus midfoot) affects the force distribution and the potential for injury. A general principle of injury mechanics suggests that as the area of force application increases, the likelihood of injury decreases. A sharp object contacting the skin with a certain amount of force, for instance, will have a different effect than a blunt object contacting the skin with the same force. The former condition might result in a puncture wound, while the latter might leave a bruise.

The measure of force and its distribution is pressure *(p)*, defined as the total applied force *(F)* divided by the area *(A)* over which the force is applied *(p = F/A)*. The standard unit of pressure, the pascal (Pa), is equal to a 1 N force applied to an area 1 m square (1 Pa = 1 N/m²). In the British system, pressure is measured in pounds per square inch (psi).

Lever Systems

Most motion at the major joints results from the body's structures acting as a system of mechanical levers. A **lever** is a rigid structure, fixed at a single point, to which two forces are applied at two different points. One of the forces is commonly referred to as the resistance force *(R)*, with the other termed the applied, or effort, force *(F)*. The fixed point, known as the axis (also fulcrum or pivot), is the point (or line) about which the lever rotates. In the human body, these three components are typically an externally applied resistance force *(R)* such as gravity, a muscle effort force *(F)*, and a joint axis of rotation *(A)*.

These three lever system components *(R, F, A)* may be spatially related to one another in three different configurations that give rise to three lever classes. Distinctions among the classes are determined by the location of each component relative to the other two (figure 6.11). In a **first-class lever**, the axis *(A)* is located between the resistance *(R)* and the effort force *(F)*. A **second-class lever** has *R* located between *F* and *A*, while a **third-class lever** has *F* between *R* and *A*. Joints in the human body are predominantly third-class levers (figure 6.12a), with some first-class levers (figure 6.12b) and few second-class levers.

Lever systems in the human body serve two important functions. First, they increase the effect of an applied force because the applied force and the resistance force usually act at different distances from the axis, as seen in the leverage advantage gained by using a long bar to

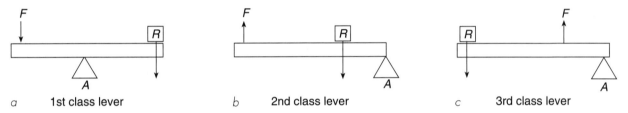

| *a* 1st class lever | *b* 2nd class lever | *c* 3rd class lever |

Figure 6.11 Lever classes: *(a)* first-class lever, *(b)* second-class lever, *(c)* third-class lever.

Reprinted, by permission, from W.C. Whiting and R.F. Zernicke, 1998, *Biomechanics of musculoskeletal injury* (Champaign, IL: Human Kinetics), 64.

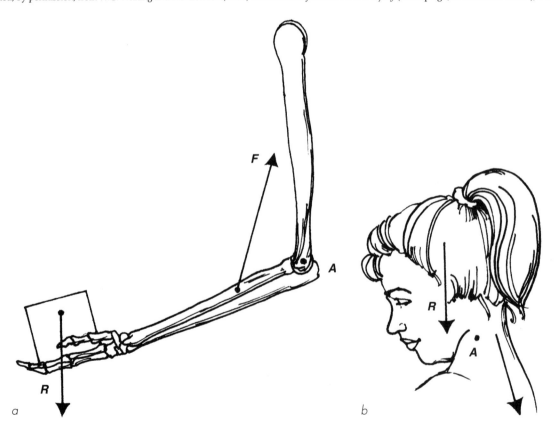

Figure 6.12 *(a)* Third-class lever system in the human body as seen in the action of the biceps at the elbow joint; *(b)* first-class lever system as seen in the head bending forward at the neck.

Reprinted, by permission, from W.C. Whiting and R.F. Zernicke, 1998, *Biomechanics of musculoskeletal injury* (Champaign, IL: Human Kinetics), 64.

pry a large rock loose from the ground. In such a first-class lever, increasing the distance from the axis to the effort force increases the effective force seen on the other side of the pivot point (i.e., it is easier to move the rock).

The second function of levers is to increase the effective speed (or velocity) of movement. During knee extension (figure 6.13), for example, a given angular displacement produces different linear displacements for points along the lower leg. Points farther away from the knee joint axis move a greater distance along the curved arc than do points closer to the axis. Because all points along the lower leg move with the same *angular* velocity, the more distal points have higher *linear* velocity. This is easily seen in baseball and volleyball. Baseball pitchers with longer arms have the potential to throw the ball faster than those with shorter arms. In volleyball, the hand speed while spiking the ball may be higher for players with longer arms.

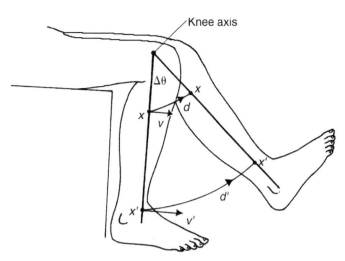

Figure 6.13   Effect of levers in increasing the speed of distal points.

Reprinted, by permission, from W.C. Whiting and R.F. Zernicke, 1998, *Biomechanics of musculoskeletal injury* (Champaign, IL: Human Kinetics), 64.

The human body makes effective use of both the force and speed advantages provided by lever systems in accomplishing the many tasks it performs on a daily basis.

Torque (Moment of Force)

In the case of linear motion, force is the mechanical agent creating and controlling movement. For angular motion, the agent is known as **torque** (*T*) (or moment of force, or **moment** (*M*)), defined as the effect of a force that tends to cause rotation or twisting about an axis (figure 6.14a).

The mathematical definition of *torque* and *moment* is the same; however, there is a technical difference between the two. Torque typically refers to the twisting movement created by a force (figure 6.14b), whereas moment is related to the bending or rotational action of a force (figure 6.14c). Despite this difference, the two terms often are used interchangeably.

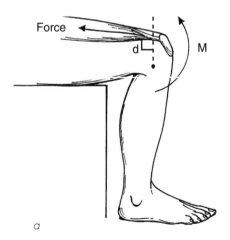

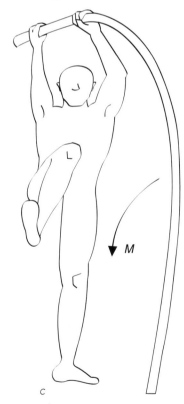

a

b

c

Figure 6.14   *(a)* Moment as the product of force and moment arm; *(b)* torque, or torsion, tending to twist a body; *(c)* moment tending to cause rotation or bending.

Reprinted, by permission, from W.C. Whiting and R.F. Zernicke, 1998, *Biomechanics of musculoskeletal injury* (Champaign, IL: Human Kinetics), 48.

The magnitude of a torque is equal to the applied force (F) multiplied by the perpendicular distance (d) from the axis of rotation to the **line of force action**. This perpendicular distance is known as the **moment arm**, **torque arm**, or **lever arm** (figure 6.15).

Biomechanical examples of torque include the biceps creating a flexion moment about the elbow, the moment created by a weight on the lower leg during a knee extension exercise, and the torque applied to the tibia in a skiing fall (figure 6.16). The standard unit of torque or moment comes from the product of the two terms: force in newtons (N) multiplied by moment arm in meters (m). The resulting unit is a newton-meter (Nm), or foot-pound (ft-lb) in the British system.

Closer examination of the moment equation (moment = force × moment arm) reveals several important general principles when applying moment concepts to movement biomechanics. First, there is an obvious interaction between the force and the moment arm that directly affects the magnitude of the applied moment. To increase the moment, we have the option of increasing the force, increasing the moment arm, or both. To decrease the moment, we can decrease either the force or moment arm, or both.

A second moment-related concept, while simple in statement, is powerful in its application. That is, when a force is applied *through* the axis of rotation, no moment is produced. This concept follows directly from the moment equation. If the force passes through the axis, the moment arm is zero, and hence no moment is produced. This can be seen by pushing on the hinges of a door. The hinges serve as the door's axis, and forces applied to the hinges will not create any moment; the door will not move.

At joints in the human body, this principle creates a situation where tissues can be exposed to extremely high forces, but no moment is created. Compressive forces acting through the center of a vertebral body, for example, will cause no vertebral rotation but may increase the risk of a compressive vertebral fracture.

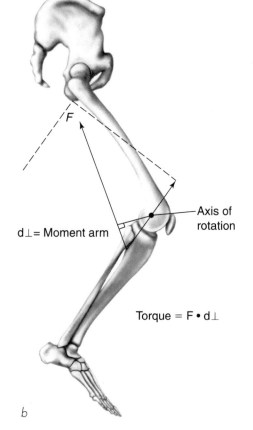

a

b

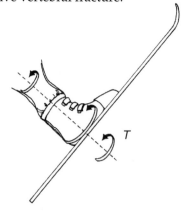

Figure 6.15 (a) Torque calculation at the knee when the line of force action is perpendicular. (b) Torque calculation when the force is not perpendicular involves using trigonometric functions.

Reprinted, by permission, from W.C. Whiting and R.F. Zernicke, 1998, *Biomechanics of musculoskeletal injury* (Champaign, IL: Human Kinetics), 50.

Figure 6.16 Torsional load being transferred from the ski and applied to the skier's lower leg.

Reprinted, by permission, from W.C. Whiting and R.F. Zernicke, 1998, *Biomechanics of musculoskeletal injury* (Champaign, IL: Human Kinetics), 50.

A third moment concept emerges from the fact that in most situations, more than one moment is being applied. The system's response is based on the **net moment** (also **net torque**) or the result of adding together all the moments acting about the given axis. An example of this concept is seen in a simple glenohumeral abduction exercise (figure 6.17). Gravity, acting on both the arm and dumbbell, creates a moment about the glenohumeral axis of rotation that tends to adduct the arm.

If this were the only moment, the arm would immediately adduct under the effect of gravity. To hold the arm in an abducted position, the abductor muscles acting about the glenohumeral joint must create a moment that acts in the opposite direction to counterbalance the moment created by gravity. This counterbalancing moment is termed a *countermoment*, or *countertorque*. The countermoment in this case will tend to abduct the arm.

The movement that results depends on the relative magnitudes of these two moments. Adding together all of the moments acting about a joint creates a net moment. If the two moments are equal in magnitude (but opposite in direction), the net moment is zero and no movement occurs. If the moment created

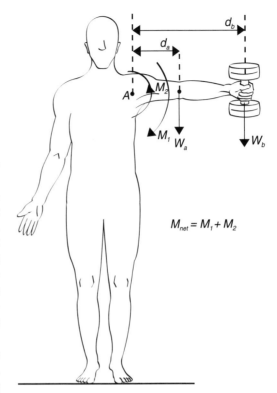

$$M_{net} = M_1 + M_2$$

Figure 6.17 Net moment, calculated as the sum of all component moments acting about a joint axis.

by gravity exceeds that created by the abductors, the net moment favors gravity and the arm will adduct. On the other hand, if the moment created by the abductors is greater than that of gravity, the net moment favors the muscle action and the arm abducts.

The importance of these moment- or torque-related concepts will become evident in forthcoming chapters as we explore the details of the production and control of human movements.

Newton's Laws of Motion

Of Sir Isaac Newton's (1642-1727) many notable scientific contributions, his laws of motion are among his most important. These three laws of motion form the basis for classical mechanics and provide the rules that govern the physics of how we move.

Newton's Laws of Motion:

- First Law of Motion (Law of Inertia): A body at rest or in motion will tend to remain at rest or in motion unless acted on by an external force.

- Second Law of Motion (Law of Acceleration): A force acting on a body will produce an acceleration proportional to the force, or mathematically, $F = m \times a$ (i.e., force = mass times acceleration).

- Third Law of Motion (Law of Action and Reaction): For every action there is an equal and opposite reaction.

Newton's three laws of motion apply to all human movements. The essence of Newton's first law is that forces are required to start, stop, or modify the movement of a body. In the absence of forces (e.g., gravity, friction), a body will persist in its state of rest or motion. For example, a skater slides across the ice until the friction between the skate and the ice eventually brings him to a stop. In another example, a dancer leaping through the air begins her flight by exerting force against the ground. Once in the air, her upward speed is slowed by the effect of

Sir Isaac Newton

Until Newton's time, the theories of Aristotle (384-322 B.C.) held sway, and scientists believed that a body's natural state was at rest and that movement required a mover, such as when a force is applied to a box to slide it across the floor. They believed that when the force stopped, the body stopped. To make this theory tenable, Aristotle and those who followed him ascribed special properties to air and water to explain the motion of bodies through fluid (gas and liquid) environments. For many centuries, scientists had no understanding of friction. Nearly two thousand years after Aristotle, Galileo (1564-1642) determined that external forces, such as friction, were required to change a body's state of motion.

Newton's *Principia Mathematica* details many of his scientific insights and consolidates and summarizes much of the work of his predecessors and contemporary colleagues, such as Copernicus, Galileo, and Descartes. In commenting on his own contributions to science and his belief that scientific knowledge is cumulative and always builds on previous work, Newton said, "If I have seen further it is by standing on the shoulders of giants" (Bartlett, 1968, p. 379). A modest statement given that Newton himself is one of the true giants of science.

gravitational forces until she reaches the peak of her jump; she then comes back to the ground under the continuing influence of gravity.

Newton's first law also is very much in evidence in the movement of astronauts in space. In the absence of gravity, the astronauts appear to "float" around the space vehicle's cabin. They actually are just continuing the motion created by the forces used to push or pull themselves relative to the cabin walls.

Common lifting tasks provide a good example of Newton's second law of motion. In lifting a box from the floor, the lifter must exert enough force to overcome the force of gravity and accelerate the box upward. The second law of motion determines the magnitude of the acceleration in response to the applied force ($F = m \times a$). For a box of a given mass (*m*), a greater force (*F*) will create a proportionally greater acceleration (*a*).

In a similar fashion, application of a moment, or torque, changes a body's angular state of motion. To accelerate the arm when throwing a ball, the muscles of the shoulder and arm must generate moments about the respective joint axes to make the arm move faster.

Newton's third law of motion states that every force action creates an equal and opposite reaction. This can be seen in a long-distance runner whose feet contact the ground many thousands of times. At each foot contact, the force that the foot exerts on the ground is equally and oppositely resisted by the ground, giving rise to the term **ground reaction force** (GRF) to describe the forces of the ground acting on the foot. Increasing the magnitude and frequency of ground reaction forces increases the chance of injury.

Equilibrium

The term **equilibrium** suggests a balanced situation. From a mechanical perspective, equilibrium exists when forces and moments (torques) are balanced. In general, equilibrium exists for a body at rest or for one moving with constant linear and angular velocities. The net forces and net moments acting on the body equal zero. In this case, the body is said to be in **static equilibrium**. Bodies in motion experiencing external forces and moments that cause acceleration are in **dynamic equilibrium**. The concepts of static and dynamic equilibrium are important in our discussion of posture and balance in chapter 8.

Work and Power

The term *work* is used in many ways, referring in various contexts to physical labor ("I'm working hard"), physiological energy expenditure ("I worked off 200 calories"), or a place of employment ("I'm going to work"). In mechanical terms, however, work has a very specific meaning. Mechanical **work** is performed by a force acting through a displacement in the direction of the force. By definition, linear work (*W*) is equal to the product of force (*F*) and the displacement (*d*) through which the body is moved (figure 6.18a). The standard (SI) unit of work is the joule (1 J = 1 Nm). If the entire force does not act in the direction of motion (figure 6.18b), only the component of force acting in the direction of movement is used in calculating the work done.

In the example depicted in figure 6.19, the work performed in lifting the barbell from point A to point B is equal to the product of the barbell's weight (W_b) and the displacement from A to B (d_{AB}). If, for example, the barbell weighs 800 N (~180 lb) and is lifted 0.5 m (~20 in.), the work done would be 400 J (800 N × 0.5 m).

The calculation of work often does not completely describe the mechanics of a body's movement or task performance. Take the case, for example, of two material handlers whose job requires them to lift boxes from the ground onto a moving conveyor belt. If the boxes each weigh 200 N (~45 lb) and the conveyor belt is 1.5 m above the ground, then 300 J of work is performed in lifting each box. However, the first lifter can lift each box and place it on the conveyor belt in 1.2 seconds, while the second lifter takes 1.4 seconds to complete the task. Because the boxes weigh the same and are moved through the same distance, each lifter performs the same amount of mechanical work. One lifter can complete the task more quickly, however, so there is some mechanical difference in how they lift. The difference is in the rate at which the work is performed. The rate of work, termed **power**, is calculated as the amount of work done divided by the time it takes to do the work (power = work ÷ time). Power is expressed in units of watts (1 W = 1 J/s).

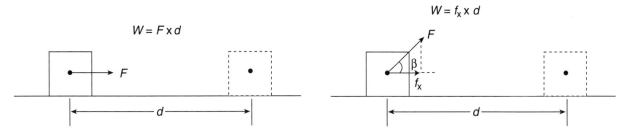

Figure 6.18   Mechanical work: (a) linear work when the force is in the direction of displacement; (b) linear work when only a portion of the force acts in the direction of displacement.

Reprinted, by permission, from W.C. Whiting and R.F. Zernicke, 1998, *Biomechanics of musculoskeletal injury* (Champaign, IL: Human Kinetics), 56.

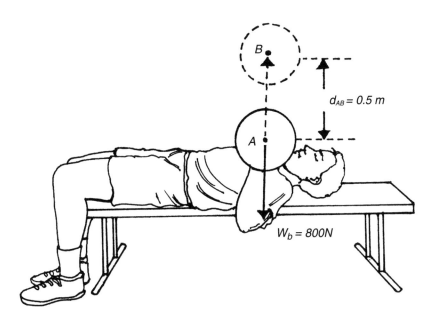

Figure 6.19   Bench press illustrating mechanical work performed.

Reprinted, by permission, from W.C. Whiting and R.F. Zernicke, 1998, *Biomechanics of musculoskeletal injury* (Champaign, IL: Human Kinetics), 56.

In the previous example, both workers perform 300 J of work. The worker who lifts a box onto the conveyor belt in 1.2 seconds lifts with a power of 250 W (300 J ÷ 1.2 s), while the worker taking 1.4 seconds lifts with a power of about 214 W (300 J ÷ 1.4 s). In general, a given amount of work performed in a shorter time yields a greater power output.

Power may also be expressed as the product of force and velocity. Successful performance of many movement tasks (e.g., jumping, throwing) requires high power output. To be powerful, the performer must generate high forces and do so quickly (i.e., high velocity). A shot-putter, for example, must be both strong (high force) and fast (high velocity) to be successful. Other examples of tasks requiring high power output are discussed in subsequent chapters.

Energy

Energy is another term with a variety of meanings. One could have a high-energy personality or, if tired, run out of energy. As with work, mechanical energy has a more specific meaning. Mechanical **energy** is defined as the capacity, or ability, to perform work. Energy can assume many forms, including thermal, chemical, nuclear, electromagnetic, and mechanical. Mechanical energy is the type most commonly used to describe or assess human movement.

The mechanical energy of a body can be classified according to its **kinetic energy** (energy of motion) or its **potential energy** (energy of position or deformation).

Linear kinetic energy (E_k) for a given body is defined as $E_k = 1/2 \times m \times v^2$, where m equals mass and v equals linear velocity of the center of mass.

Angular kinetic energy $(E_{\angle k})$ is defined as $E_{\angle k} = 1/2 \times I \times \omega^2$, where I equals mass moment of inertia and ω equals angular velocity. Energy is measured in joules (J), the same units used to measure work.

In terms of human movement, one of the most important concepts to emerge from the kinetic energy equations is that the velocity term $(v$ or $\omega)$ is squared. As movement speed increases, energy multiplies as a function of the velocity squared. For example, consider a downhill skier with a body mass of 60 kg who increases her downhill velocity from 20 m/s to 25 m/s. Her velocity has increased by 25%, but her linear kinetic energy has increased by more than 56%. What might at first appear to be a relatively small increase in velocity considerably enhances her kinetic energy. And if the skier unexpectedly (and unfortunately) collides with a tree or another skier, the higher energy can result in serious injury.

Potential energy can take two forms. The gravitational form (potential energy of position) measures the potential to perform work as a function of the height a body is elevated above some reference level, most typically the ground. The equation describing **gravitational potential energy** is $E_p = m \times g \times h$, where m equals mass, g equals gravitational acceleration (~9.81 m/s²), and h equals height (in meters) above the reference level.

The second form of potential energy is **deformational** (also called **strain**) **energy** that is stored in a body when it is deformed. The deformation can take the form of a body being stretched, compressed, bent, or twisted. Common examples of strain energy include a stretched rubber band or tendon, a pole-vaulter's bent pole, a drawn bow before arrow release, and a compressed intervertebral disc. When the force causing the deformation is removed, some of the stored strain energy is returned to use, while the rest is lost as heat energy.

Scientists studying whole-body or limb segment movement dynamics often assume that each body segment is rigid (i.e., nondeformable). When this simplifying assumption is made, there is no strain energy component in the system. In these cases, the **total mechanical energy** is simply the sum of the linear kinetic, angular kinetic, and positional potential energies. (Note that total mechanical energy also includes a thermal energy term, but because it usually is negligible compared with the other terms, we omit it here.)

Thus, total mechanical energy (TME) equals linear kinetic energy plus angular kinetic energy plus positional potential energy. Consider, for example, a soccer player swinging her leg to kick the ball. Each of the lower-limb segments (thigh, shank, foot) has a continuously changing total mechanical energy. Let's focus on the shank (lower leg) to illustrate how to measure the total mechanical energy. As the values in figure 6.20 show, at the peak of the leg's backswing, the kinetic energy is essentially zero because the shank is almost motionless, and the positional potential energy is at its highest. As the leg swings forward toward the ball, the potential energy decreases (i.e., the shank is lowered), and the kinetic energy increases as the leg swings faster. At contact, much of the leg's energy is transferred to the ball, and the ball accelerates toward its target. More details of kicking dynamics are presented in chapter 9.

Momentum

In an athletic contest, when one team is playing well and seems to be getting all of the breaks, we say that the players have momentum on their side. From a mechanical perspective, however, **momentum** is defined differently and can be characterized as a measure of a body's "quantity of motion." In general, the larger the body and the faster it is going, the higher its momentum. In mechanical terms, **linear momentum** is the product of mass *(m)* and velocity *(v)*. Increasing either a body's size (mass) or speed (velocity) will increase its linear momentum.

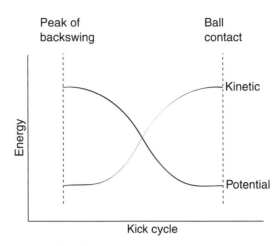

Figure 6.20   Components of energy as seen during a kicking motion.

Similarly, **angular momentum**, or the "quantity of angular motion," is the product of mass moment of inertia *(I)* and angular velocity *(ω)*.

Principles of Conservation and Transfer

Two principles govern the effect of both energy and momentum on human movement: conservation and transfer. First, let's look at how these principles apply to momentum. **Conservation of momentum** indicates how much of a system's combined linear and angular momentum, or quantity of motion, is conserved and how much is gained or lost during a given time period. This principle is a consequence of Newton's first law of motion, which implies that a body's momentum is conserved (i.e., remains the same) unless the body changes its mass or is acted on by an external force.

A person running along the beach, for example, has some amount of momentum based on his body size (mass) and how fast he is running (velocity). If he stops, he loses all of his momentum because his velocity drops to zero. In contrast, when a sprinter is in the starting blocks awaiting the starter's gun, her momentum is zero because she is motionless. When the gun sounds, she pushes against the starting blocks and accelerates. As her velocity increases, so does her momentum. In neither of these cases is momentum conserved. In the first case, momentum is lost; in the second, momentum is gained.

The companion principle, **transfer of momentum**, is the mechanism by which momentum from one body is transferred to another. This can take many forms in human movement. Transfer during a throwing motion can occur as momentum moves from a proximal segment (e.g., upper arm) to a more distal segment (e.g., forearm, hand) as the throw progresses.

Momentum can also be transferred between different bodies, as in the case of an automobile crash or an American football player blocking or tackling an opponent. Transfer of momentum in these cases often results in injury when the quantity of motion transferred exceeds the tolerance of the tissues in one or both of the bodies.

The principles of conservation and transfer can also be applied to energy. Conservation of energy dictates that a body's total energy is conserved (i.e., remains constant) unless the body changes its mass or is acted on by an outside force. Transfer of energy, in a manner similar to transfer of momentum, explains how energy is passed from one body to another.

Friction

Newton's first law of motion tells us that bodies in motion tend to remain in motion unless acted on by an outside force. That force may be an abrupt one, such as a collision, or of lower magnitude and greater duration, such as the force of friction. **Friction** is defined as the resistance created at the interface between two bodies in contact with one another and acting in a direction opposite to impending or actual movement. Frictional resistance results from microscopic irregularities, known as asperities, on the opposing surfaces. Asperities tend to adhere to each other, and efforts to move the bodies result in very small resistive forces that oppose the motion.

In the simple case of a body at rest on a surface, **static friction** resists movement until a force sufficient to overcome the frictional resistance is applied. As the force applied to a body at rest increases, a level is reached at which the static resistance is overcome, and the body begins to slide along the surface. Once the body begins moving, the friction decreases slightly and then is known as **kinetic friction**.

If the tires of a car are prevented from rotating (e.g., when one slams on the brakes), the vehicle slides along the road. In this case, kinetic friction (as just described) is in effect. Most of the time, however, the wheels are free to rotate and the vehicle rolls forward. Even in rolling movement, friction is present. Rolling resistance is not as obvious as sliding resistance because rolling friction is much lower, often by a factor of 100 to 1,000 times. The actual value of resistance in both sliding and rolling friction depends on the material properties of the body and the surface and on forces acting between them.

In some cases friction works to our advantage. In fact, we would be unable to walk or run without friction acting between our shoes and the ground. Too little friction, such as when someone walks on an icy surface, may lead to a slip and fall. Too much friction, on the other hand, may contribute to injuries of another type. High levels of friction lead to abrupt deceleration, which causes high forces and extreme loading of body tissues.

Our examples thus far have focused on friction acting externally on the body. Friction also plays an important role within the human body. During normal limb movements, for example, the friction in synovial joints is extremely low, allowing for freedom of movement with minimal resistance.

Fluid Mechanics

Fluid mechanics, the branch of mechanics dealing with the properties and behavior of gases and liquids, plays an important role in our consideration of human movement. Areas as diverse as performance biomechanics (study of movement mechanics), tissue biomechanics (study of the mechanical response of tissues), and hemodynamics (study of blood circulation) all rely on the basic principles of fluid mechanics.

We move in various fluid environments, where air is the principal gas and water is the predominant liquid (e.g., swimming). We consider two important mechanical properties: flow and resistance. Fluid flow refers to the characteristics of a fluid, whether liquid or gas, that allow it to move and govern the nature of this movement. Blood flow through an artery is an example of fluid flow. Fluids also provide resistance, such as the resistance we might experience while running against the wind or swimming in a pool. Understanding fluid flow and resistance is essential to our understanding of human movement.

Fluid Flow

Fluid flow can exhibit many movement patterns. **Laminar flow** is characterized by a smooth, essentially parallel pattern of movement, as seen in the waters of a slow-moving river. **Turbulent flow**, in contrast, exhibits a more chaotic flow pattern, characterized by areas of turbulence and multidirectional fluid movement. Arterial blood flow provides a good example of these different types of flow. In the middle of the artery, blood flow may be laminar. Blood in contact with the arterial wall or at a branching in the artery may exhibit turbulence. Factors contributing to turbulent flow include the roughness of the surface over which the fluid passes, the diameter of the vessel, obstructions, and speed of flow.

Fluid Resistance

Fluid resistance takes many forms, some of which are advantageous to human movement and others that may prove detrimental. Examples of positive effects of fluid resistance include **buoyant forces**, which allow a person to float in water; **aerodynamic forces**, which keep an object in flight; and **magnus forces**, which affect the trajectories of objects spinning through the air. Negative effects of fluid resistance are evident in the extra physiological effort required by a cyclist moving into the wind or by the severe and unpredictable forces acting on a hang glider during a storm.

Resistance to flow is termed **viscosity**. Sometimes described as "fluid friction," viscosity enables a fluid to develop and maintain a resistance to flow, dependent on the flow's velocity. The effect of viscosity and its dependence on velocity can be seen in a familiar example. When you move your hand slowly through water, the resistance is minimal. Increasing the speed of movement markedly increases the resistance. The increase in resistance is due to the liquid's viscosity-related, or viscous, properties.

Fluid mechanical effects play an integral role in human movement tasks such as swimming. Specific application of fluid mechanics principles to swimming is presented in chapter 10.

Joint Mechanics

Human movement relies on the action of hundreds of articulations (joints) in the human body. As discussed in earlier chapters, the amount of movement depends on each joint's structure. No two joints are the same in structural terms; each has its own distinct combination of tissues, tissue configuration, and movement potential. This variety of joint structure and function allows for complex movement patterns.

Each joint in the body has an operational range of motion (ROM) that determines the joint's mobility. The allowable ROM is both joint specific and person specific. Joints with an ability to move in more than one plane have ROMs specific to each particular plane of movement. Note that ROMs vary considerably from one person to another, and thus individual measurement is the surest method of determining accurate joint ROMs. Refer to table 3.4 (page 53) for average ranges of joint motion for specific joints. Integrally related to ROM is the notion of joint stability, defined as "the ability of a joint to maintain an appropriate functional position throughout its range of motion" (Burstein & Wright, 1994, p. 63).

One way of viewing joint stability is to consider the joint's ability to resist dislocation. Stable joints have a high resistance to dislocation. Unstable joints tend to dislocate more easily. By way of review from chapter 3, recall that in general, joints can be classified along a stability–mobility continuum, which specifies that joints with a tight bony fit and numerous ligamentous and other supporting structures or surrounded by large muscle groups will be very stable and relatively immobile. Joints with a loose bony fit, limited extrinsic support, or minimal surrounding musculature tend to be very mobile and unstable. The classic exception to this categorization is the hip joint, which is both very mobile—with large ROM potential in all three primary planes—and very stable, as evidenced by the rarity of its dislocation.

Material Mechanics

So far our discussion has focused primarily on the movement of bodies and the external forces affecting those movements. In this section, we shift our attention briefly to the human body's internal mechanics, focusing on the internal response of tissues to externally applied loads. Although not directly related to movement mechanics, fundamental principles of **material mechanics** do influence the integrity of musculoskeletal structures and their susceptibility to injury. Injury-related tissue damage can compromise movements in many ways. Thus, an understanding of the basics of material mechanics is essential to our study of human movement.

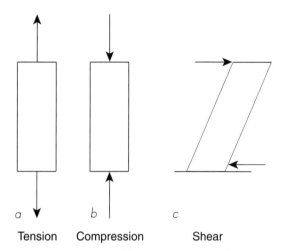

Tension   Compression   Shear

Figure 6.21 Types of loads: *(a)* tension, *(b)* compression, *(c)* shear.

Properties of biological materials (e.g., tissues) influence a material's response to forces. Among these properties are size, shape, area, volume, and mass. The tissue's structural constituents and form, discussed in earlier chapters, also play an important factor in its mechanical response.

Externally applied forces, termed **loads**, come in three types: tension, compression, and shear. Tensile (**tension**) loads tend to pull ends apart (figure 6.21a). Compressive (**compression**) loads tend to push ends together (figure 6.21b). **Shear** loads tend to produce horizontal sliding of one layer over another (figure 6.21c).

Stress and Strain

Tissues mechanically loaded by an external force develop an internal resistance to the load. In the case of a thin rubber band, this resistance is minimal. In contrast, the resistance developed by a steel rod is considerable. This internal resistance to loading, called mechanical **stress**, is common to all materials. Materials change their shape, though sometimes imperceptibly, in response to external loads. This change in shape is termed deformation, or mechanical **strain**.

A direct relationship exists between stress and strain, and the consequences of this relationship in a tissue determine its response to loading. Bone loaded by compression, for example, develops high resistance while deforming very little. Skin, in contrast, deforms considerably more at substantially lower forces. The stress–strain responses of tendons, ligaments, and cartilage fall somewhere between those of bone and skin.

The stress–strain relationship can be summarized in a single measure as the ratio of the two values. This stress–strain ratio defines a material's **stiffness**. The opposite, or inverse, of stiffness is known as **compliance**. Stiff materials, such as bone, have high stress–strain ratios. More compliant materials, such as skin, have lower ratios.

Bending and Torsion

In our earlier discussion of moment of inertia, we briefly mentioned the resistance to both bending and twisting (or torsion). Any relatively long and slender structure (e.g., a long bone) may be considered in mechanical terms as a beam. Any force or moment tends to bend the beam.

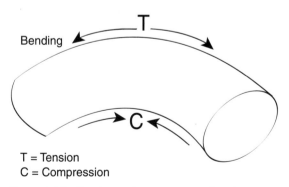

T = Tension
C = Compression

Figure 6.22 Bending with tension created on the convex surface and compression developed on the concave surface.

In bending, the material on the concave (inner) surface of the structure experiences compressive stress, while that on the convex (outer) surface is subject to tensile stress (figure 6.22). These tensile and compressive stresses are maximal at the outer surfaces of the beam and diminish toward the middle of the beam.

Any twisting action applied to a structure results in torsional loading, as seen in the simple example of unscrewing a lid from its jar. Angular resistance is involved when torsional loads are applied to a body. For example, when a skier's leg is twisted, his tibia experiences torsional loading (or torque). The internal stresses developed in response to the torsional loading produce resistance to the applied torque.

Viscoelasticity

As noted at the beginning of our discussion of material mechanics, the mechanical response of a material depends on its constituent matter, which in the case of biological tissues includes a fluid component (e.g., water). A tissue's viscosity provides resistance to flow and makes the mechanical response of the tissue partially dependent on the rate at which it is loaded. As a result, we say that the tissue's mechanical response is **strain-rate dependent**, which means that the response depends on the rate at which the tissue is deformed. A tendon stretched quickly, for example, will be weaker and stiffer than the same tendon stretched more slowly. This strain-rate dependency is important because tissue strain rate is related to the speed of movement. Rapid movements elicit different tissue responses than do slower movements.

Another important tissue property is elasticity, defined as a tissue's ability to return to its original shape, or configuration, when a load is removed. Biological tissues, largely by virtue of their fluid content, have combined viscous and elastic responses. We therefore often describe tissues as having **viscoelasticity**. The viscoelastic properties of tissues play an important role in their mechanical response and indirectly can have profound effects on movement. A rapidly stretched tendon, for example, can store strain energy, some of which is released as the tendon subsequently shortens. This return of stored energy can improve movement performance. A person performing a vertical jump can jump higher, for instance, if she rapidly stretches the musculotendinous structures in the leg by performing a quick squat immediately before jumping upward. More examples of how tissue properties affect human movements are discussed in later chapters.

concluding comments

Many of the mechanical concepts presented in this chapter are evident as we explore human movement in the following chapters. Keep in mind that although the information here was divided into separate sections for presentation, the concepts interrelate in complex and fascinating ways. One of the essential challenges in studying the biomechanics of human movement is to identify the salient aspects of movement, fluid, joint, and material mechanics and blend them into an understandable and useful whole.

critical thinking questions

1. Define center of mass or center of gravity, and describe its relative importance in analyzing human movement. Use a specific exercise or movement to accompany your discussion. During normal standing, where would you expect to find your body's center of mass, and during movement, does it always remain inside the body? Use a specific example to support this last question.

2. Define base of support. Use a specific exercise or movement to show how you can change your base of support to either increase or decrease your stability and how these changes can affect performance.

3. Define and describe the difference between inertia and moment of inertia. Include a brief discussion as to why these are important concepts when analyzing human movement. In addition, define the three different types of moment of inertia and provide specific applications.

4. Define and compare or contrast the following terms: *force*, *pressure*, *torque*, *work*, *power* and *energy*. Use equations where appropriate, and provide specific exercise or movement applications for each term.

5. Define Newton's three laws of motion, and provide specific performance applications for each one.

6. Define potential and kinetic energy, and give their equations. Be sure to include both the linear and angular equations for kinetic energy. In addition, compare and contrast kinetic energy and momentum. Demonstrate how you would apply all these terms if you were analyzing a specific sport.

7. What is the difference between static and kinetic friction? Use specific applications.

8. With respect to material mechanics, describe the three principal types of loads.

9. Briefly discuss the direct relationship between stress and strain. Provide specific applications for each term.

10. Define viscoelasticity, and provide a specific example of how a tissue's viscoelastic properties can be used to enhance performance.

suggested readings

Enoka, R.M. (2002). *Neuromechanics of human movement* (3rd ed.). Champaign, IL: Human Kinetics.

Hall, S.J. (2003). *Basic biomechanics* (4th ed.). New York: McGraw-Hill.

McGinnis, P.M. (2005). *Biomechanics of sport and exercise* (2nd ed.). Champaign, IL: Human Kinetics.

Neumann, D.A. (2002). *Kinesiology of the musculoskeletal system: Foundations for physical rehabilitation.* St. Louis: Mosby.

Nordin, M., & Frankel, V.H. (2001). *Basic biomechanics of the musculoskeletal system* (3rd ed.). Philadelphia: Lippincott Williams & Wilkins.

seven

Muscular Control of Movement and Movement Assessment

objectives

After studying this chapter, you will be able to do the following:

- Describe concepts of muscle function: agonist action, neutralization, stabilization, antagonist action, and coactivation

- Use the muscle control formula to determine muscle action for any movement

- Explain movement concepts of coordination, efficiency, and economy

- Describe sources of movement inefficiency

- Describe methods of kinematic, kinetic, and electromyographic movement assessment

> A man cannot think deeply and exert his utmost muscular force.
> —Charles Darwin (1809-1882)

As we learned in chapter 4, the amount of force generated by a given muscle is influenced by many factors, including the nature of the task, speed of movement, amount of external resistance, level of activation, and muscle length and velocity. In addition, the force needed by the muscle must be determined in the context of other muscles' actions and the constraints imposed by the principles of mechanics and muscle physiology.

In most movements, mechanical and physiological variables change continuously throughout the movement and thus make the nervous system's job complex. Although the hows and whys of muscle control remain somewhat a mystery, the fact that the neuromuscular system *does* control our actions is beyond dispute. Does it do so perfectly? Hardly. But in most cases, our bodies do a commendable job of movement control.

The story of muscular control of movement is analogous to the children's story of the three bears. The moral of that story is that things shouldn't be "too hot" or "too cold" but should be "just right." The same rule applies to muscle action. Too much or too little force compromises movement control. In the words of futurist Alvin Toffler, "Overcontrol is just as dangerous as undercontrol" (Toffler, 1990, p. 463).

At times, generating too much force can be just as detrimental to performance as not producing enough force, as in the case of someone just learning to perform a task. Novices, in their unfamiliarity with a particular movement, may excessively recruit muscles and create hesitant, jerky, and inefficient movements. Skilled performers, in contrast, make good use of their muscles and execute confident, fluid, and coordinated movements. In terms of movement control, sometimes less is more.

Too much conscious thought can compromise the effectiveness of movement, a principle sometimes jokingly referred to as "paralysis by analysis." In matters of the mind and its effect on movement control, noted Russian actor and director Konstantin Stanislavsky (1863-1938)

In Others' Words: William James

In one sense, the more or less of tension in our faces and in our unused muscles *is* a small thing; not much mechanical work is done by these contractions. But it is not always the material size of a thing that measures its importance, often it is its place and function.

One of the most philosophical remarks I ever heard was by an unlettered workman who was doing some repairs at my house many years ago. 'There is very little difference between one man and another,' he said, 'when you go to the bottom of it. But what little there is, is very important.'

And the remark certainly applies to this case. The general over-contraction may be small when estimated in foot-pounds, but its importance is immense on account of its effects on the over-contracted person's spiritual life.

This follows as a necessary consequence from the theory of emotions, . . . [f]or by the sensations that so incessantly pour in from the over-tense excited body the over-tense and excited habit of mind is kept up; and the sultry, threatening, exhausting, thunderous inner atmosphere never quite clears away.

If you never wholly give yourself up to the chair you sit in, but always keep your leg and body muscles half contracted for a rise; if you breathe eighteen or nineteen instead of sixteen times a minute, and never quite breathe out at that—what mental mood *can* you be in but one of inner panting and expectancy, and how can the future and its worries possibly forsake your mind? On the other hand, how can they gain admission to your mind if your brow be unruffled, your respiration calm, and complete and your muscles all relaxed? (James, 1983, p.122-123)

—William James (1842-1910), noted psychologist and philosopher

observed, "At times of great stress, it is especially necessary to achieve a complete freeing of the muscles" (Stanislavski, 1948, p. 94). Similar sentiments have been expressed by many others as well (see one other view in the box on page 120).

Muscle Function

The control of even the simplest joint movement typically requires the cooperative action of several muscles working together as a single unit. This cooperative action is called **muscle synergy**. Synergistic muscles work together, but other muscles with opposite functions may work against a particular movement. The overall, or net, effect of all muscles acting at a joint determines the ultimate mechanical effect, or movement.

Several concepts of muscle function are important for understanding how muscles cooperate and compete to control movement: agonists, neutralization, stabilization, antagonists, and coactivation.

- *Agonists:* Muscles that actively control a single joint movement or maintain a single joint position are called **agonists**. In most movements, several muscles act together as agonists, with some playing a greater role than others.

- *Neutralization:* Muscles often perform more than one movement function at a given joint. At the ankle complex, for example, a muscle might act as both a plantar flexor and invertor. To produce pure plantar flexion, another muscle whose action produces plantar flexion and eversion would also need to be involved. The eversion action of the second muscle would cancel out, or neutralize, the inversion action of the first muscle. This process of canceling out an unwanted secondary movement is called **neutralization**.

- *Stabilization:* During concentric action, a muscle attempts to shorten by pulling its two bony attachment sites together. In most cases, the bone with the least resistance to movement (inertia) will move. When the inertia of both bones is similar in magnitude, both ends tend to move. If movement of only one end is desired, the other end must be prevented from moving, or stabilized. This **stabilization** is provided by other muscles or an external force. As an example, consider hip flexion created by the anterior thigh musculature. In attempting to move the femur in flexion, the hip flexors also tilt the pelvis anteriorly. If pelvic tilt is unwanted, then the abdominal musculature must act isometrically to stabilize the pelvis and prevent its movement.

- *Antagonists:* Muscles acting against a movement or position are called **antagonists**. To perform a movement most effectively, when the agonists actively shorten in concentric action, the corresponding antagonists *passively lengthen.* When agonists actively lengthen in eccentric action, the associated antagonists *passively shorten.* In many movements, then, the agonists and stabilizers are active while the antagonists are passive.

- *Coactivation:* Simultaneous action of both agonists *and* antagonists is called **coactivation**, or *co-contraction.* Coactivation might occur, for example, when an unskilled performer is unsure of the necessary muscle recruitment strategy. Skilled performers, however, do not exhibit an absence of coactivation. There are three possible explanations for coactivation in skilled performers: (1) less overall effort may be required in agonist–antagonist pairings for movements that involve changes of direction when the muscles maintain some level of activity, as opposed to working in an on–off manner; (2) coactivation increases joint stiffness and consequently joint stability, which may be desired for movements involving heavy loads; and (3) coactivation of a single-joint muscle (e.g., gluteus maximus) and a two-joint muscle (e.g., rectus femoris) can increase the torque at a joint (e.g., knee) acted on by the two-joint muscle (Enoka, 1994). Note that the term *coactivation* is limited to the concurrent action of agonists and antagonists and should *not* be used to describe the simultaneous action of multiple agonists. For instance, concurrent activity of the biceps brachii and triceps brachii during an elbow curl exercise would be considered coactivation. On the other hand, if the triceps was passive (i.e., inactive), simultaneous action of the three elbow flexors (biceps brachii, brachialis, brachioradialis) would *not* be considered coactivation.

With these concepts in mind, we now consider a simple yet fundamental question: How do we determine which muscles are active in producing or controlling a given movement?

Muscle Action

One of the most fundamental and important goals of movement analysis is identifying which specific muscles are active in producing and controlling movement at a particular joint. In chapter 5, we presented specific muscles and their *concentric* actions. As presented in chapter 4, however, we know that muscles can act in three modes: isometric, concentric, eccentric. The task at hand, therefore, is to determine for a given joint movement (1) the specific muscles involved in controlling the movement and (2) the type of muscle action.

The following **muscle control formula** provides a step-by-step procedure for determining the involved muscles and their action for any joint movement. This formula may appear a bit cumbersome and complex at first glance. But with practice, you should be able to get through it quickly. Eventually (with enough practice), the process will become automatic and instinctive, and you will be able to analyze movements without consciously going through each step in the formula. It helps, though, to use the formula until you develop these movement analysis instincts.

Muscle Control Formula

We begin the muscle control formula with a statement of the problem: Given a specific joint movement (or position), identify the name of the movement (or position), the plane of movement, the effect of the external force acting on the system, the type of muscle action (i.e., shortening/concentric, lengthening/eccentric, or isometric), and the muscle(s) involved (i.e., which muscle or muscles are actively involved in producing or controlling the movement or in maintaining a position).

Now, we move on to the formula itself, which involves six steps:

Step 1
Identify the *joint movement* (e.g., flexion, abduction) or position.
Step 2
Identify the *effect of the external force* (e.g., gravity) on the joint movement or position by asking the following question: "What movement would the external force produce in the absence of muscle action (i.e., if there were no active muscles)?"
Step 3
Identify the *type of muscle action* (concentric, eccentric, isometric) based on the answers to step 1 (#1) and step 2 (#2) as follows:

   a. If #1 and #2 are in *opposite* directions, then the muscles are actively shortening in a *concentric action*. Speed of movement is *not* a factor.

   b. If #1 and #2 are in the *same* direction, then ask yourself, "What is the speed of movement?"

      i. If the movement is *faster* than what the external force would produce by itself, then the muscles are actively shortening in a *concentric action*.

      ii. If the movement is *slower* than what the external force would produce by itself, then the muscles are actively lengthening in an *eccentric action*.

   c. If no movement is occurring, yet the external force would produce movement if acting by itself, then the muscles are performing an *isometric action*.

   d. Movements *across gravity* (i.e., parallel to the ground) are produced by a *concentric action*. When gravity cannot influence the joint movement in question, shortening (concentric) action is needed to pull the bone against its own inertia. The speed of movement is *not* a factor.

By this point, we have identified the type of muscle action. The next steps identify which muscles control the movement.

Step 4

Identify the *plane of movement* (frontal, sagittal, transverse) and the *axis of rotation* (i.e., line about which the joint is rotating). The purpose of this step is to identify which side of the joint the muscles controlling the movement cross (e.g., flexors cross one side of a joint, while extensors cross the opposite side).

Step 5

Ask yourself, "On which side of the joint axis are muscles lengthening and on which side are they shortening during the movement?"

Step 6

Combine the information from steps 3 and 5 to *determine which muscles must be producing or controlling the movement* (or position). For example, if a concentric (shortening) action is required (from step 3) and the muscles on the anterior side of the joint are shortening (from step 5), then the anterior muscles must be actively producing the movement. The information in chapter 5 allows us to name the specific muscles.

Application of the Muscle Control Formula

Let's see how to apply the muscle control formula by going through the step-by-step process for some simple single-joint movements. For all these examples, we assume *no* coactivation of antagonists, which allows us to simplify the analysis by eliminating the need to consider the effect of antagonist muscles. Coactivation of agonist–antagonist groups stiffens the joint and makes joint motion more difficult.

Example 1: Biceps Curl

Consider the simple movement (figure 7.1) of elbow flexion as the person moves the joint from position A (elbow fully extended) to position B (elbow flexed).

Step 1: The movement is flexion.

Step 2: The external force (gravity) tends to extend the elbow.

Step 3: The movement (flexion) is opposite that created by the external force, so the muscles are actively shortening in *concentric action.*

Step 4: Movement occurs in the sagittal plane about an axis through the elbow joint.

Step 5: The muscles on the joint's anterior surface are shortening during the movement, while the muscles on the posterior side are lengthening.

Step 6: The muscle action is concentric (from step 3), and muscles on the anterior side of the joint are short-

Figure 7.1 Biceps curl: elbow flexion from position A to B; elbow extension from position B to A.

ening (from step 5). Thus, the anterior muscles actively produce the movement. Using the information from chapter 5, we know that the biceps brachii, brachialis, and brachioradialis are the muscles responsible for the movement.

Now consider the movement (figure 7.1) of elbow extension as the person moves the joint from position B (elbow flexed) to position A (elbow fully extended). The movement speed is slow, meaning the movement occurs *slower* than what the external force would produce acting by itself (i.e., in the absence of muscle action).

Step 1: The movement is extension.

Step 2: The external force (gravity) tends to extend the elbow.

Step 3: The movement (extension) is the same as that created by the external force, so ask yourself, "What is the speed of movement?" The speed is slow, which dictates that the controlling muscles are actively lengthening in an *eccentric action.*

Step 4: Movement occurs in the sagittal plane about an axis through the elbow joint.

Step 5: The muscles on the joint's anterior surface are lengthening, while the muscles on the posterior side are shortening.

Step 6: The muscle action is eccentric (from step 3), and muscles on the anterior surface are lengthening (from step 5). Thus, the anterior muscles actively control the movement. Again, using the information from chapter 5, we identify the biceps brachii, brachialis, and brachioradialis as the muscles responsible for the movement.

In this example, we see that the muscles normally identified as elbow flexors (i.e., biceps brachii, brachialis, brachioradialis) act concentrically to produce elbow flexion and also act eccentrically to control elbow extension. This principle that "flexors" can control extension may, at first, seem counterintuitive. Why wouldn't the so-called elbow extensors (e.g., triceps brachii) control elbow extension? The answer lies in the speed of the movement. Movements that are slower than what the external force would create (in the absence of muscle action) are controlled by eccentric action of muscles on the side opposite of those that would produce the movement concentrically. This principle of eccentric control cannot be overemphasized.

Now consider the same movement (figure 7.1) of elbow extension as the person moves from position B (elbow flexed) to position A (elbow fully extended) but this time by moving fast, meaning that the movement occurs *faster* than what the external force would produce acting by itself (i.e., in the absence of muscle action). In this action, the arm is rapidly "snapped" into extension.

Step 1: The movement is extension.

Step 2: The external force (gravity) tends to extend the elbow.

Step 3: The movement (extension) is the same as that created by the external force, so ask yourself, "What is the speed of movement?" The speed is fast, which dictates that the controlling muscles are actively shortening in a *concentric action.*

Step 4: Movement occurs in the sagittal plane about an axis through the elbow joint.

Step 5: The muscles on the joint's anterior surface are lengthening during the movement, while the muscles on the posterior side are shortening.

Step 6: The muscle action is concentric (step 3), and muscles on the posterior side of the joint are shortening (step 5). Thus, the posterior muscles actively produce the movement. Therefore, the elbow extensors, primarily the triceps brachii, are responsible for the movement.

It is important to note that in this case, the speed of movement (fast) determines that the movement (elbow extension) is produced by concentric action of the elbow extensors. In the previous case, the speed of movement (slow) dictates that eccentric action of the elbow flexors is required to control joint extension.

Example 2: Leg Extension

Consider the leg extension movement shown in figure 7.2. The knee joint is extended from position A (knee flexed) to position B (knee fully extended).

Step 1: The movement is extension.

Step 2: The external force (gravity) tends to flex the knee.

Step 3: The movement (extension) is opposite that created by the external force, so the muscles are actively shortening in *concentric action*.

Step 4: Movement occurs in the sagittal plane about an axis through the knee joint.

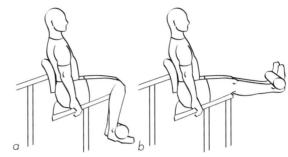

Figure 7.2 Leg extension: knee extension from position A to B; knee flexion from position B to A.

Step 5: The muscles on the joint's anterior surface are shortening during the movement, while the muscles on the posterior side are lengthening.

Step 6: The muscle action is concentric (step 3), and muscles on the anterior side of the joint are shortening (step 5). Thus, the anterior muscles actively produce the movement. Therefore, the muscles of the quadriceps group (vastus medialis, vastus lateralis, vastus intermedius, rectus femoris) produce the movement.

Now consider the movement (figure 7.2) of knee flexion as the joint moves from position B (knee extended) to position A (knee flexed). The movement speed is slow, meaning the movement occurs *slower* than what the external force would produce acting by itself (i.e., in the absence of muscle action).

Step 1: The movement is flexion.

Step 2: The external force (gravity) tends to flex the knee.

Step 3: The movement (flexion) is the same as that created by the external force, so ask yourself, "What is the speed of movement?" The speed is slow, which dictates that the controlling muscles are actively lengthening in an *eccentric action*.

Step 4: Movement occurs in the sagittal plane about an axis through the knee joint.

Step 5: The muscles on the joint's anterior surface are lengthening, while the muscles on the posterior side are shortening.

Step 6: The muscle action is eccentric (step 3), and muscles on the anterior surface are lengthening (step 5). Thus, the anterior muscles actively control the movement. Therefore, the vastus medialis, vastus lateralis, vastus intermedius, and rectus femoris control the movement.

Consider the same movement (figure 7.2) of knee flexion as the joint moves from position B (knee extended) to position A (knee flexed) but this time moving fast, meaning the movement occurs *faster* than what the external force would produce acting by itself (i.e., in the absence of muscle action). In this action, the leg is rapidly "snapped" into flexion.

Step 1: The movement is flexion.

Step 2: The external force (gravity) tends to flex the knee.

Step 3: The movement (flexion) is the same as that created by the external force, so ask yourself, "What is the speed of movement?" The speed is fast, which dictates that the controlling muscles are actively shortening in a *concentric action*.

Step 4: Movement occurs in the sagittal plane about an axis through the knee joint.

Step 5: The muscles on the joint's anterior surface are lengthening during the movement, while the muscles on the posterior side are shortening.

Step 6: The muscle action is concentric (step 3), and muscles on the posterior side of the joint are shortening (step 5). Thus, the posterior muscles actively produce the movement. Therefore, the knee flexors (e.g., biceps femoris, semitendinosus, semimembranosus) produce the movement.

Example 3: Dumbbell Lateral Raise

This example involves raising the arms from the side of the body (position A) to parallel with the ground (position B) as shown in figure 7.3.

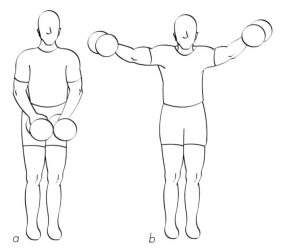

a *b*

Figure 7.3 Dumbbell lateral raise: arm abduction from position A to B; arm adduction from position B to A.

Step 1: The movement is abduction.

Step 2: The external force (gravity) tends to adduct the arm.

Step 3: The movement (abduction) is opposite that created by the external force, so the muscles are actively shortening in *concentric action.*

Step 4: Movement occurs in the frontal plane about an axis through the glenohumeral joint.

Step 5: The muscles on the joint's superior surface are shortening during the movement, while the muscles on the inferior side are lengthening.

Step 6: The muscle action is concentric (step 3), and muscles on the superior side of the joint are shortening (step 5).

Thus, the superior muscles (the abductors) actively produce the movement. Therefore, the anterior and middle deltoid and supraspinatus produce the movement.

Now consider the movement (figure 7.3) of glenohumeral adduction as the person moves the joint from position B (abducted) to position A (adducted), this time slowly.

Step 1: The movement is adduction.

Step 2: The external force (gravity) tends to adduct the arm.

Step 3: The movement (adduction) is the same as that created by the external force, so ask yourself, "What is the speed of movement?" The speed is slow, which dictates that the controlling muscles are actively lengthening in an *eccentric action.*

Step 4: Movement occurs in the frontal plane about an axis through the glenohumeral joint.

Step 5: The muscles on the joint's superior surface are lengthening, while the muscles on the inferior side are shortening.

Step 6: The muscle action is eccentric (step 3), and muscles on the superior surface are lengthening (step 5). Thus, the superior muscles actively control the movement. Therefore, the deltoid and supraspinatus control the movement.

Consider the same movement (figure 7.3) of glenohumeral adduction, this time moving fast. In this action, the arm is rapidly "snapped" to the side of the body from its abducted position.

Step 1: The movement is adduction.

Step 2: The external force (gravity) tends to adduct the arm.

Step 3: The movement (adduction) is the same as that created by the external force, so ask yourself, "What is the speed of movement?" The speed is fast, which dictates that the controlling muscles are actively shortening in a *concentric action.*

Step 4: Movement occurs in the frontal plane about an axis through the glenohumeral joint.

Step 5: The muscles on the joint's superior surface are lengthening during the movement, while the muscles on the inferior side are shortening.

Step 6: The muscle action is concentric (step 3), and muscles on the inferior side of the joint are shortening (step 5). Thus, the inferior muscles actively produce the movement. Therefore, the glenohumeral adductors (e.g., pectoralis major, latissimus dorsi, teres major) produce the movement.

Example 4: Abdominal Curl-Up

When performing a curl-up exercise, the exerciser flexes the trunk from position A (trunk fully extended) to position B (trunk flexed) as shown in figure 7.4.

Step 1: The movement is flexion.

Step 2: The external force (gravity) tends to extend the trunk.

Step 3: The movement (flexion) is opposite that created by the external force, so the muscles are actively shortening in *concentric action.*

Step 4: Movement occurs in the sagittal plane about multiple axes through the different vertebrae of the spine.

a

Step 5: The muscles on the joint's anterior surface are shortening during the movement, while the muscles on the posterior side are lengthening.

Step 6: The muscle action is concentric (step 3), and muscles on the anterior side of the joint are shortening (step 5). Thus, the anterior muscles actively produce the movement. Therefore, the rectus abdominis and the internal and external obliques produce the movement.

b

Figure 7.4 Abdominal curl-up: trunk flexion from position A to B; trunk extension from position B to A.

To return to the starting position, the exerciser extends the trunk from position B (trunk flexed) to position A (trunk fully extended) in a slow and controlled manner.

Step 1: The movement is extension.

Step 2: The external force (gravity) tends to extend the trunk.

Step 3: The movement (extension) is the same as that created by the external force, so ask yourself, "What is the speed of movement?" The speed is slow, which dictates that the controlling muscles are actively lengthening in an *eccentric action.*

Step 4: Movement occurs in the sagittal plane about multiple axes through the different vertebrae of the spine.

Step 5: The muscles on the joint's anterior surface are lengthening, while the muscles on the posterior side are shortening.

Step 6: The muscle action is eccentric (step 3), and muscles on the anterior surface are lengthening (step 5). Thus, the anterior muscles actively control the movement. Therefore, the rectus abdominis, assisted by the internal and external obliques, again control the movement.

This analysis of muscle action indicates that curl-up (and related) exercises require almost continuous abdominal muscle activity, concentrically during the up phase and eccentrically during the controlled down phase. "Snapping" the trunk back to the ground at a speed faster than gravity would be both difficult and inadvisable because of the risk of injury. We therefore do not analyze that condition.

Example 5: Standing Calf Raise

A common exercise for strengthening the calf muscles on the posterior aspect of the lower leg (shank) involves isolated plantar flexion of the ankle, often with added resistance in the form of carried weights or a weight machine designed for this purpose. This exercise (commonly called heel raises or calf raises) is typically performed with the ball of the foot on a raised surface to allow the rear of the foot to drop lower on the down phase. Consider first the movement from position A (ankle in neutral anatomical position) to position B (ankle plantar flexed) as shown in figure 7.5.

Figure 7.5 Standing calf raise: ankle plantar flexion from position A to B; ankle dorsiflexion from position B to A.

Step 1: The movement is plantar flexion.

Step 2: The external force (gravity) tends to dorsiflex the ankle.

Step 3: The movement (plantar flexion) is opposite that created by the external force, so the muscles are actively shortening in *concentric action*.

Step 4: Movement occurs in the sagittal plane about an axis through the ankle joint.

Step 5: The muscles on the joint's posterior surface are shortening during the movement, while the muscles on the anterior side are lengthening.

Step 6: The muscle action is concentric (step 3), and muscles on the posterior side of the joint are shortening (step 5). Thus, the posterior muscles actively produce the movement. Therefore, the soleus and gastrocnemius, with slight assistance by the other plantar flexors (peroneus longus, peroneus brevis, tibialis posterior, flexor hallucis longus, flexor digitorum longus, plantaris), produce the movement.

During the lowering phase of the movement, the ankle *slowly* moves from position B (ankle plantar flexed) to position A (ankle neutral).

Step 1: The movement is dorsiflexion.

Step 2: The external force (gravity) tends to dorsiflex the ankle.

Step 3: The movement (dorsiflexion) is the same as that created by the external force, so ask yourself, "What is the speed of movement?" The speed is slow, which dictates that the controlling muscles are actively lengthening in an *eccentric action*.

Step 4: Movement occurs in the sagittal plane about an axis through the ankle joint.

Step 5: The muscles on the joint's posterior surface are lengthening, while the muscles on the anterior side are shortening.

Step 6: The muscle action is eccentric (step 3), and muscles on the posterior surface are lengthening (step 5). Thus, the posterior muscles actively control the movement. Therefore, the soleus and gastrocnemius (again with assistance from the other plantar flexors) control the movement.

Example 6: Shoulder Rotation

In the examples considered so far, gravity has been the external force. Consider now an example where the external force is other than gravity, as might be the case of an astronaut using elastic (bungee) cords in the microgravity of outer space or a patient undergoing physical therapy who uses elastic bands for resistance. In this example, the subject is trying to strengthen rotator muscles acting at the glenohumeral joint. Working against the resistance provided by an elastic band (figure 7.6), the subject moves his arm from position A (glenohumeral joint externally rotated) to position B (internally rotated).

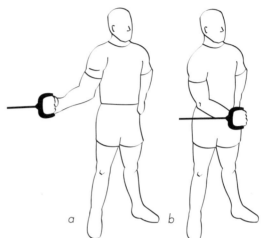

Step 1: The movement is internal rotation.

Step 2: The external force (elastic band) tends to externally rotate the gleno-humeral joint.

Step 3: The movement (internal rotation) is opposite that created by the external force, so the muscles are actively shortening in *concentric action*.

Step 4: Movement occurs in the transverse plane about a vertical axis through the glenohumeral joint.

Step 5: The muscles on the joint's anterior surface are shortening during the movement, while the muscles on the posterior side are lengthening.

Figure 7.6 Shoulder rotation: glenohumeral internal rotation from position A to B; glenohumeral external rotation from position B to A.

Step 6: The muscle action is concentric (step 3), and muscles on the anterior side of the joint are shortening (step 5). Thus, the anterior muscles actively produce the movement. Therefore, the subscapularis, pectoralis major, anterior deltoid, latissimus dorsi, and teres major produce the movement.

If the subject *slowly* returns from position B to the starting position A, the muscle action is determined as follows:

Step 1: The movement is external rotation.

Step 2: The external force (elastic band) tends to externally rotate the glenohumeral joint.

Step 3: The movement (external rotation) is the same as that created by the external force, so ask yourself, "What is the speed of movement?" The speed is slow, which dictates that the controlling muscles are actively lengthening in an *eccentric action*.

Step 4: Movement occurs in the transverse plane about a vertical axis through the glenohumeral joint.

Step 5: The muscles on the joint's anterior surface are lengthening, while the muscles on the posterior side are shortening.

Step 6: The muscle action is eccentric (step 3), and muscles on the anterior surface are lengthening (step 5). Thus, the anterior muscles actively control the movement. Therefore, the same internal rotators (subscapularis, pectoralis major, latissimus dorsi, anterior deltoid, teres major) control the movement.

If the external rotation was performed quickly, the muscles would cause external rotation faster than the elastic band could develop tension. The muscles basically would defeat the purpose of the elastic band in providing resistance, and the band would become slack. The muscle action in this case would be determined as follows:

Step 1: The movement is external rotation.

Step 2: The external force (elastic band) tends to externally rotate the glenohumeral joint.

Step 3: The movement (external rotation) is the same as that created by the external force, so ask yourself, "What is the speed of movement?" The speed is fast, which dictates that the controlling muscles are actively shortening in a *concentric action*.

Step 4: Movement occurs in the transverse plane about a vertical axis through the glenohumeral joint.

Step 5: The muscles on the joint's anterior surface are lengthening during the movement, while the muscles on the posterior side are shortening.

Step 6: The muscle action is concentric (step 3), and muscles on the posterior side of the joint are shortening (step 5). Thus, the posterior muscles actively produce the movement. Therefore, the external rotators (infraspinatus, teres minor, posterior deltoid) produce the movement.

Placing the elastic band on the other side (i.e., so that it pulls from a medial direction) reverses the muscle actions. The external rotators (infraspinatus, teres minor, posterior deltoid) would act concentrically to externally rotate the arm against the elastic band's resistance, which tends to internally rotate the joint. In returning to the starting (internally rotated) position slowly, the same external rotators would act eccentrically to control the internal rotation.

Example 7: Standing Dumbbell Flys

Some movements occur parallel to the ground and, therefore, aren't directly affected by gravity. Consider an exercise (figure 7.7) in which the exerciser holds light dumbbells in an abducted position (A). Note that although muscle action is required to keep the arms in an elevated position, movements in the transverse plane are not directly affected by gravity. The muscle control formula is applied as follows:

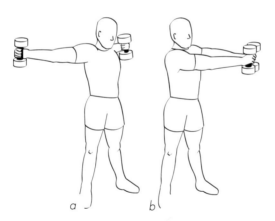

Figure 7.7 Standing dumbbell flys: glenohumeral horizontal adduction from position A to B; glenohumeral horizontal abduction from position B to A.

Step 1: The movement is horizontal adduction (also called horizontal flexion).

Step 2: No external force affects this motion because the movement is parallel to the ground (no effect of gravity for movement in the transverse plane), and there are no other external forces (e.g., elastic bands).

Step 3: The movement (horizontal adduction) occurs *across gravity,* so the muscles are actively shortening in *concentric action* to overcome the inertia of the limb segments.

Step 4: Movement occurs in the transverse plane about a vertical axis through the glenohumeral joint.

Step 5: The muscles on the shoulder joint's anterior surface are shortening during the movement, while the muscles on the posterior side are lengthening.

Step 6: The muscle action is concentric (step 3), and muscles on the anterior side of the joint are shortening (step 5). Thus the anterior muscles actively produce the movement. Therefore, the pectoralis major and anterior deltoid primarily produce the movement.

In returning from position B to the starting position A (figure 7.7), the muscle action is determined as follows:

Step 1: The movement is horizontal abduction (also called horizontal extension).

Step 2: No external force affects this motion because the movement is parallel to the ground (no effect of gravity), and there are no other external forces (e.g., elastic bands).

Step 3: The movement (horizontal abduction) occurs *across gravity,* so the muscles are actively shortening in *concentric action* to overcome the inertia of the limb segments.

Step 4: Movement occurs in the transverse plane about a vertical axis through the glenohumeral joint.

Step 5: The muscles on the joint's posterior surface are shortening during the movement, while the muscles on the anterior side are lengthening.

Step 6: The muscle action is concentric (step 3), and muscles on the posterior side of the joint are shortening (step 5). Thus, the posterior muscles actively produce the movement. Therefore, the posterior deltoid, teres minor, infraspinatus, teres major, and latissimus dorsi primarily produce the movement.

Example 8: Back Extension

Let's consider one final example in this section to cover the case when no motion occurs, but muscle force is needed to maintain joint position. What muscle action is required for the person in figure 7.8 to hold the trunk in the position shown? (Note: This exercise can place high loads on the spine and so is not recommended for general use. It is presented here only to show how the muscle control formula can be used to determine muscle function.)

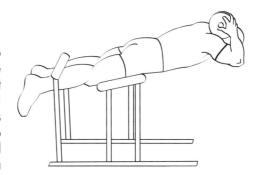

Figure 7.8 Back extension: trunk held in neutral position while suspended above the ground.

Step 1: The trunk's position is fully extended (i.e., anatomical position).

Step 2: The external force (gravity) tends to flex the trunk.

Step 3: No movement is occurring, yet the external force would produce movement if allowed to act by itself, so the muscles are performing an *isometric action.*

Step 4: If movement was allowed, gravity would flex the trunk in the sagittal plane about multiple axes through the different vertebrae of the spine.

Step 5: The muscles on the joint's anterior surface would shorten during the movement (if allowed to occur), while the muscles on the posterior side would lengthen.

Step 6: The muscle action is isometric (step 3). To prevent muscle length changes on both sides of the trunk (step 5), the posterior muscles (erector spinae) must be active to prevent trunk flexion and counteract the action of gravity.

Single-Joint Versus Multijoint Movements

Our examples thus far have involved movement at a single joint. We now consider two multijoint movements to show how multiple applications of the muscle control formula can fully describe muscle actions in more complex multijoint movements. More examples of multijoint movements are discussed in chapters 8, 9, and 10.

Example 9: Squat

Many activities of daily living, as well as many athletic movements, require us to move from an upright standing position into some form of a squat. A typical squatting movement is shown in figure 7.9 as the person moves from position A (upright standing) to position B (squat). This movement invariably is done *slowly.* The squat is a multijoint movement involving primarily the hip, knee, and ankle joints.

As the exerciser moves from position A to position B, the muscle action is determined as follows:

Figure 7.9   Squat: downward phase from position A to B; upward phase from position B to A.

Step 1: The movements are hip and knee flexion and ankle dorsiflexion.

Step 2: The external force (gravity) tends to flex the hip and knee and dorsiflex the ankle.

Step 3: The movements are the same as those created by the external force, so ask yourself, "What is the speed of movement?" The speed is slow, which dictates that the controlling muscles are actively lengthening in an *eccentric action* at all three joints.

Step 4: Movement occurs in the sagittal plane about axes through the hip, knee, and ankle joints.

Step 5: The muscles on the hip joint's anterior surface are shortening, while the muscles on the posterior side are lengthening. At the knee, the anterior muscles are lengthening, and the posterior muscles are shortening. At the ankle, the anterior muscles are shortening, and the posterior muscles are lengthening.

Step 6: The muscle action is eccentric (step 3). Muscles on the posterior surface of the hip and ankle are lengthening (step 5). At the knee, muscles on the joint's anterior surface are lengthening (step 5). Therefore, the downward phase of the squat is controlled by eccentric action of the hip extensors (gluteus maximus, semitendinosus, semimembranosus, long head of the biceps femoris, posterior fibers of the adductor magnus), knee extensors (vastus medialis, vastus lateralis, vastus intermedius, rectus femoris), and ankle plantar flexors (soleus and gastrocnemius, with slight assistance of the peroneus longus and brevis, tibialis posterior, flexor hallucis longus, flexor digitorum longus, and plantaris).

In returning from the squatting position (B) to upright standing (position A), the muscle action is determined as follows:

Step 1: The movements are hip and knee extension and ankle plantar flexion.

Step 2: The external force (gravity) tends to flex the hip and knee and dorsiflex the ankle.

Step 3: The movements are opposite those created by the external force, so the muscles are actively shortening in *concentric action* at all three joints.

Step 4: Movement occurs in the sagittal plane about axes through the hip, knee, and ankle joints.

Step 5: The muscles on the hip joint's anterior surface are lengthening, while the muscles on the posterior side are shortening. At the knee, the anterior muscles are shortening, and the posterior muscles are lengthening. At the ankle, the anterior muscles are lengthening, and the posterior muscles are shortening.

Step 6: The muscle action is concentric (step 3). Muscles on the posterior surface of the hip and ankle are shortening (step 5). At the knee, muscles on the joint's anterior surface (step 5) are shortening. Therefore, the upward phase of the squat is produced by concentric action of the hip extensors (gluteus maximus, semitendinosus, semimembranosus, long head of the biceps femoris, posterior fibers of the adductor magnus), knee extensors (vastus medialis, vastus lateralis, vastus intermedius, rectus femoris), and ankle plantar flexors (soleus and gastrocnemius, with slight assistance of the peroneus longus and brevis, tibialis posterior, flexor hallucis longus, flexor digitorum longus, and plantaris).

You may have noticed in several of the examples that eccentric action of muscles is immediately followed by concentric action of the same muscles. This pattern of eccentric–concentric coupling is the mechanism underlying the stretch–shorten cycle (explained in chapter 4). Additional functional applications of the stretch–shorten cycle (e.g., vertical jumping) are presented and explained in subsequent chapters.

Example 10: Overhead Press

In some situations it is necessary to lift, or press, an object from shoulder level to an overhead position. In a warehouse, for example, workers often lift boxes to place them on a high shelf. In the gym, a common exercise is an overhead press using a barbell, as shown in figure 7.10. The pressing movement involves simultaneous actions of the shoulder girdle and glenohumeral and elbow joints. The exerciser begins with the barbell at chest level (position A) and then presses the barbell overhead to position B. For this example, we restrict movement to the frontal plane, and analyze only actions at the glenohumeral and elbow joints.

Step 1: The movements are glenohumeral abduction and elbow extension.

Step 2: The external force (gravity) tends to adduct the glenohumeral joint and flex the elbow.

Step 3: The movements are opposite those created by the external force, so the muscles are actively shortening in *concentric action.*

Step 4: Movement occurs in the frontal plane about axes through the glenohumeral and elbow joints.

Figure 7.10 Overhead press: upward phase from position A to B; downward phase from position B to A.

Step 5: The muscles on the glenohumeral joint's superior surface are shortening during the movement, while the muscles on the inferior side are lengthening. At the elbow, the posterior muscles are shortening, and the anterior muscles are lengthening.

Step 6: The muscle action is concentric (step 3). Muscles on the superior surface of the glenohumeral joint and posterior surface of the elbow are shortening (step 5). Therefore, the upward phase of the overhead press is produced by concentric action of the glenohumeral abductors (supraspinatus, anterior deltoid, middle deltoid) and elbow extensors (primarily triceps brachii).

In lowering the bar from position B back to the starting position A in a slow and controlled manner, the muscle action is determined as follows:

Step 1: The movements are glenohumeral adduction and elbow flexion.

Step 2: The external force (gravity) tends to adduct the glenohumeral joint and flex the elbow.

Step 3: The movements are the same as those created by the external force, so ask yourself, "What is the speed of movement?" The speed is slow, which dictates that the controlling muscles are actively lengthening in an *eccentric action* at both joints.

Step 4: Movement occurs in the frontal plane about axes through the glenohumeral and elbow joints.

Step 5: The muscles on the glenohumeral joint's inferior surface are shortening during the movement, while the muscles on the superior side are lengthening. At the elbow, the anterior muscles are shortening, and the posterior muscles are lengthening.

Step 6: The muscle action is eccentric (step 3). Muscles on the superior surface of the glenohumeral joint and posterior surface of the elbow are lengthening (step 5). Therefore, the downward phase of the overhead press is controlled by eccentric action of the glenohumeral abductors (supraspinatus, anterior deltoid, middle deltoid) and elbow extensors (primarily triceps brachii).

For the purpose of illustration only, consider the same movement, this time moving fast. In this action, the arms and barbell are rapidly "snapped" back to the chest. This would *not* be a recommended movement because of the risk of injury.

Step 1: The movements are glenohumeral adduction and elbow flexion.

Step 2: The external force (gravity) tends to adduct the glenohumeral joint and flex the elbow.

Step 3: The movements are the same as those created by the external force, so ask yourself, "What is the speed of movement?" The speed is fast, which dictates that the controlling muscles are actively shortening in a *concentric action* at both joints.

Step 4: Movement occurs in the frontal plane about axes through the glenohumeral and elbow joints.

Step 5: The muscles on the glenohumeral joint's inferior surface are shortening during the movement, while the muscles on the superior side are lengthening. At the elbow, the anterior muscles are shortening, and the posterior muscles are lengthening.

Step 6: The muscle action is concentric (step 3). Muscles on the inferior surface of the glenohumeral joint and anterior surface of the elbow are shortening (step 5). Therefore, rapid movement during the downward phase of the overhead press would be produced by concentric action of the glenohumeral adductors (pectoralis major, latissimus dorsi, teres major) and elbow flexors (biceps brachii, brachialis, brachioradialis).

Coordination of Movement

A skilled mover is often described as being coordinated. We, of course, have an intuitive idea of what this means, but what exactly is coordination? In general terms, **coordination** is "the harmonious functioning of parts for effective results" (Mish, 1997, p. 255). This general definition actually fits movement coordination well. Only through the harmonious functioning of our body's parts (anatomical structures and systems) can we achieve effective results (smooth and effective movement). More specifically, coordination requires that various muscles work together with correct timing and intensity to produce or control a movement. Muscular coordination is required for all movements, from simple to complex. Skilled performers of all kinds have a special ability to recruit the appropriate muscles, at the right time and with the right level of activation, to produce and control movement of individual segments and the body as a whole. Subsequent chapters provide many examples of coordinated actions controlled by the body's neuromuscular system.

Movement Efficiency

In biomechanical terms, **efficiency** refers to how much mechanical output (work) is produced for a given amount of metabolic input (energy). The ratio of mechanical output to metabolic input describes the efficiency of a process. Muscle, for example, has an efficiency of about 25%, meaning that only one-quarter of the metabolic energy goes toward performing work, while the other three-quarters is converted to heat or used in energy recovery processes (Enoka, 1994). Efficient movements are characterized by relatively high work output with low metabolic energy expenditure.

A related measure, movement **economy**, is not synonymous with efficiency. Economy refers to how much metabolic energy is required to perform a given amount of work. In brief, efficiency applies to constant-energy conditions (i.e., how much work can be done using a given amount of energy), while economy corresponds with constant-work situations (i.e., how much energy is required to perform a constant amount of work) (Enoka, 1994). The relationship between efficiency and economy is shown in figure 7.11.

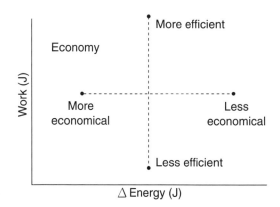

Figure 7.11 The relationship between efficiency and economy.

In most instances, movement efficiency is desirable. Common sense dictates that we want to produce as much work as possible with a minimum metabolic energy expenditure. In some cases, however, efficiency might not be the highest priority. When confronted with a dangerous situation, for example, the objective might be to move as quickly as possible or to generate as much force as possible, with little or no consideration for efficiency. In general, however, movement efficiency can be considered an important goal, and inefficiencies detract from the effectiveness of movement.

Several actions or conditions can contribute to movement inefficiency:

• *Muscular coactivation:* The net torque required at a joint to perform a given movement is determined by the sum of the torques created by all the muscles acting at that joint. The most efficient way to generate a needed torque is to activate muscles on only one side of the joint (i.e., no coactivation of muscles on the opposite side). Antagonistic coactivation works against the action of agonist muscles attempting to move the joint and, strictly speaking, contributes to inefficiency. Keep in mind, though, that coactivation can provide certain performance advantages (as explained earlier).

• *Jerky movements:* Movements characterized by rapid starts and stops or by alternating changes of direction are inefficient because metabolic energy is required to decelerate and accelerate the segments. If mechanical work is determined by the overall movement from one point to another, then jerky movements performed in getting there contribute to inefficiency. For example, the gait of a child with cerebral palsy may include spastic limb movements that require considerable metabolic energy expenditure, thereby increasing the effort required to walk from one place to another.

• *Extraneous movements:* Nonessential movements that do not directly contribute to completing a movement task nonetheless use metabolic energy and thus are inefficient. A runner, for instance, who adopts a peculiar running style that involves swinging his arms in a windmill fashion would run inefficiently because circumduction of the arms is extraneous to running. Some arm movement certainly is required to assist in balance and to run effectively, but excessive arm movement reduces efficiency.

• *Isometric actions:* As described in chapter 6, mechanical work is calculated as the product of force and displacement. Because isometric muscle actions involve no movement, they produce no mechanical work. Thus, the metabolic energy expended in producing isometric force is "wasted" in terms of producing mechanical work but may be necessary for stability.

• *Large center of gravity excursions:* Metabolic energy is required to raise and lower the body's center of gravity. In walking, for example, some vertical oscillation of the center of gravity is necessary when moving horizontally from one point to another. Excessive up and down motion of the center of gravity, however, is superfluous to the intended task and thus makes the movement less efficient.

Muscle Redundancy

Detailed exploration of the many mechanical and physiological factors involved in determining how much force a muscle needs to generate is beyond the scope of this text (see the suggested readings), but presentation of one such factor will give you an idea of the complexity of the nervous system's task.

Consider a simple uniplanar hinge joint such as the elbow. To flex this joint, how many muscles are *minimally* required? One, of course. Most joints in the body, however, have more than one muscle crossing each side of the joint. At the elbow, for example, three muscles (biceps brachii, brachialis, brachioradialis) perform flexion. Having more than the minimally required number of muscles creates what is termed **muscle redundancy**. This redundancy presents the nervous system with a challenging problem: in a task that requires less than maximal effort of all involved muscles, how does the nervous system decide how much force each muscle should provide?

Consider the case of a simple elbow curl (flexion) exercise with a 10 lb dumbbell. For most people, this is a submaximal task, meaning that none of the three elbow flexors is maximally activated in flexing the joint. Curling the dumbbell requires a certain net moment, or torque. Each of the three elbow flexors supplies a portion of this moment. Each muscle's moment is determined by the product of its respective force and moment arm, as described in the previous chapter. So how does the nervous system determine how much force will be provided by the biceps brachii, brachialis, and brachioradialis? Is the responsibility divided neatly into thirds, with each muscle making an equal contribution? Does each muscle make the same contribution during each repetition of the exercise? As the muscles fatigue, how does the division of labor change to meet the continuing demands of the task? These questions are not easily answered. And the questions become all the more complex when considering the dynamics of multijoint movements. This area of nervous system decision making remains one of the primary and continuing challenges in human movement studies.

Fortunately, from a practical standpoint, we don't need to worry much about *how* the nervous system accomplishes these complex tasks. In most cases, our neural circuitry handles the challenge without our conscious involvement. The body takes into account many variables such as muscle fiber length, cross-sectional area, attachment sites and angles of pull, fiber type, and joint angle, to name but a few, and decides how much force each muscle contributes. Remarkably, the body's neuromuscular system processes all these details with little conscious effort.

Movement Assessment

We all assess movement. Some do it professionally as clinicians, therapists, teachers, coaches, or exercise scientists. Others informally assess movement just by observing other humans moving, whether it is watching a baby's first steps, a child at play, or the slow and crouched gait of an 80-year-old. Most movement assessment is made qualitatively, as when a coach evaluates her athletes and corrects their sport technique. In some instances, however, precise quantitative measures are needed to make fine distinctions between movements. Exercise scientists and clinicians are among those who need the information provided by such quantitative analyses to diagnose a problem, assess the causes of movement dysfunction, improve performance, or increase the safety of a particular movement.

Quantitative analysis usually falls under the domain of biomechanics, the interdiscipline that applies mechanical principles to the study of biological organisms and systems. Professionals trained in biomechanics have the tools to perform detailed quantitative assessment of many aspects of human movement. Some studies measure kinematics (i.e., movement description without regard to the forces and torques involved) of a particular movement pattern. Typically, more complex studies measure the kinetic (i.e., force-related) factors involved in performing a task or movement pattern. Some of the tools used to measure the kinematics and kinetics of human movement are described in this section.

Kinematic Assessment

Substantive quantitative evaluation of movement was impossible until appropriate technologies were developed in the 19th and 20th centuries. Photography, developed in the 1830s, permitted the capture of still images from which measurements of position could be made. Nineteenth-century scientists, such as the influential French scientist Etienne-Jules Marey, used photographic techniques (along with other technologies) to capture human motion. Pioneers such as Marey laid the groundwork for the development of cinematography, or motion pictures. Although most people associate motion pictures with movies made for entertainment, it is interesting to note that *scientific* cinematography actually predates entertainment cinematography.

Who invented motion pictures? Historical evidence paints an unclear picture, with various people given credit. Unquestionably, though, two men played an important role. The first was Marey. The other was Eadweard Muybridge, a noted photographer of the late 1800s. Muybridge's photographic collections on human and animal locomotion are classics in the field.

Muybridge published countless panels of sequential still photographs that depict many kinds of movements. With proper equipment, such image sequences could be projected rapidly in succession and transformed into motion pictures. Muybridge's role in this development has an interesting (see accompanying box) and important history. His contributions, while lacking scientific rigor, are enduring. "The richness of his recordings testify to the importance that the language of images, and particularly the language of moving images, was to have in scientific research, documentation, and modern communication" (Tosi, 1992, p. 57).

Cinematography proved an indispensable instrument for the observation, recording, and evaluation of human movement during much of the 20th century. In recent decades, cinematography has been replaced, in many instances, by videographic technologies. Although some laboratories still use film for image capture and movement analysis, the many advantages of videotape (e.g., lower cost, no need to develop film) have made videography the dominant medium in recent years.

Sophisticated computerized analysis systems now can provide detailed assessment of human movement. Current systems are based on one of several different technologies. Each technology has its own advantages and limitations, and researchers must determine which system best suits their particular needs. Because some systems are expensive, available resources often dictate the type of system used.

Many systems employ videography to perform frame-by-frame analysis of images captured by standard video cameras or by more sophisticated (and expensive) high-speed video cameras.

Muybridge and the Stanford Experiment

In the late 1800s, horse racing was both a popular sport and social event. Among its many devotees was Leland Stanford, a wealthy U.S. industrialist. Story has it that Stanford, who was aware of Muybridge's work, commissioned him to provide photographic evidence to settle a purported wager as to whether all four of a horse's hooves were off the ground at any point during its running cycle. Muybridge set up a series of cameras on Stanford's farm in Palo Alto, California, and confirmed that horses do have a flight phase when no hooves are in contact with the ground. As an historical side note, the farm where Muybridge took his photographs later became the site of the now internationally recognized Stanford University.

Standard video cameras capture sequential images at a rate of approximately 30 frames per second (fps). Some analysis systems can take these 30 fps images and "split" them electronically to yield an effective frame rate of 60 fps. This speed is adequate for analyzing many, but not all, human movements. Walking and weightlifting, for example, can be reasonably evaluated at a 60 fps frame rate.

Faster movements, such as throwing or kicking, or ballistic (i.e., explosive) sports movements (e.g., swinging a golf club or baseball bat) require higher frame rates. To analyze these movements effectively, a movement scientist needs a camera capable of frame rates of 120 fps, 240 fps, or even higher.

Frame-by-frame video images are analyzed with the assistance of a computerized system in a process called **digitization**. Each image is *digitized* to identify the specific location (coordinates) of anatomical landmarks of interest. The digitized coordinates then can be used to describe the mechanical characteristics (kinematics) of the movements. For example, to describe the lower-extremity movements of a child with cerebral palsy, a clinical researcher might be interested in the movement patterns of the hip, knee, and ankle joints. To quantify these patterns, the locations of joint centers would be digitized and the resulting information used to describe the joint movements.

New technologies have emerged that allow for sophisticated movement analysis (figure 7.12). The need for movement analysis crosses many scientific, industrial, and entertainment areas. Clinicians use motion analysis systems to assess the movement patterns of special populations (e.g., stroke patients) to better identify appropriate treatment interventions and rehabilitation programs; sport scientists use movement analysis to identify the nuances of movement patterns in elite athletes in hopes of finding ways to improve performance and minimize the risk of injury; and ergonomics experts employ motion analysis to assess movements in the workplace to improve worker productivity and efficiency.

In recent years, the most visible application of motion analysis has been in the entertainment industry. Rapid advances in imaging and computer technology have afforded animation artists, cinematographers, and video-game developers the tools to produce remarkably lifelike dynamic images. Experts in these areas use sophisticated *motion capture* technologies to collect detailed data on the movement patterns of human performers and use these data to replicate movements in animated form. The astonishing rate of technological advancement in the areas of motion capture and computer animation foretells even more remarkable movement imagery in the future.

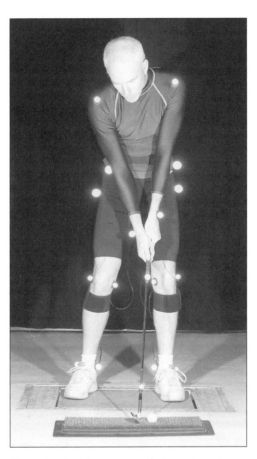

Figure 7.12 Movement analysis equipment.

Reprinted, by permission, from P. McGinnis, 2004, *Biomechanics of sport and exercise*, 2nd edition, (Champaign, IL: Human Kinetics), 361.

Kinetic Assessment

Descriptive information (kinematics) sometimes is all that is needed to answer the questions or solve the problems of movement analysts. In some cases, however, it is also important to know the causes or mechanisms underlying the observed movements. The force-related (kinetic) variables involved in movement are measured in a variety of ways.

Two of the most commonly used tools to measure force are force plates and isokinetic dynamometers. A force plate (figure 7.13) is a rigid metal platform with embedded sensors that measure the force applied to the plate. Using the principle of Newton's third law of motion, the force plate measures the equal and opposite forces generated between the body (usually through the foot) and the ground. These measured ground reaction forces are useful in understanding the loads the body experiences during certain activities and can be integrated into complex mathematical models to predict the forces and torques (moments) acting at major joints in the body (e.g., ankle, knee, hip) during tasks such as walking, running, and jumping.

Isokinetic dynamometry is widely used in both clinical and research settings (figure 7.14). Dynamometers measure kinetic characteristics of isolated joint motions (e.g., knee flexion and extension) at specific and nominally constant angular velocities. Isokinetic devices can measure parameters such as net joint torque, work, and power and are used to assess muscle strength in controlled, dynamic situations. Isokinetic dynamometry has been widely used over the past quarter century across many areas of study, including rehabilitation and sports medicine.

Electromyography

In chapter 4, we discussed the electrical characteristics of the excitation–contraction coupling processes of skeletal muscle and how we measure the electrical activity developed by muscle during the force generating process using electromyography (EMG). Electromyography is used across a wide variety of disciplines ranging from basic research on skeletal muscle mechanics to application in occupational, sport, and clinical settings.

As a clinical tool, EMG is useful in assessing the characteristics of normal voluntary muscle activation and comparing this with the activity patterns of patients with muscle weakness, paralysis, spasticity, or neuromuscular lesions. EMG is also used clinically to measure nerve conduction velocities. *Clinical electromyography* provides physicians and clinicians with valuable information to assist in diagnosis, treatment, and rehabilitation of neuromuscular disorders.

Electromyography also has application in assessing the coordination (timing and activation level) of muscles during movement tasks. *Kinesiological electromyography,* as this area is called, employs EMG to measure muscle activity patterns in sport,

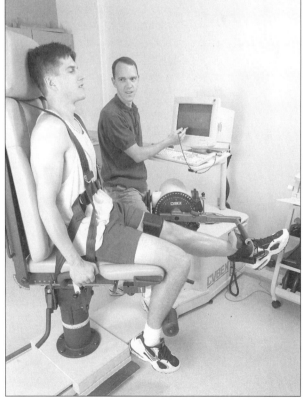

Figure 7.13   Force platform, or force plate.

Reprinted, by permission, from P. McGinnis, 2004, *Biomechanics of sport and exercise,* 2nd edition, (Champaign, IL: Human Kinetics), 361.

Figure 7.14   Isokinetic dynamometer to measure joint strength and speed.

occupational, and rehabilitation settings to assist in identifying appropriate training programs for performance enhancement.

Electromyography as an area of study traces its roots to the work of Italian anatomist Luigi Galvani (1737-1798), who in the late 18th century observed electrical activity in the muscles of frogs' legs. Advances in electromyographic instrumentation and experimentation, and thereby in our understanding of neuromuscular physiology and mechanics, progressed rapidly in the 19th and 20th centuries. Among the notable scientists engaged in electromyographic research were early pioneers such as Emil Du Bois-Reymond and C.B. Duchenne du Boulogne. Much of what we have learned about the functional aspects of movement control using EMG is summarized in the classic work *Muscles Alive* (Basmajian & DeLuca, 1985).

Many movement-specific electromyographic studies are presented in subsequent chapters as we explore fundamental human movements; movement dysfunction in clinical practice; and specialized areas of exercise, sport, and dance.

concluding comments

Our muscles miraculously control movement with little conscious effort. The nervous system controls these body engines and calls on them when needed to perform all of our daily movement tasks. In a simple lifting task, for example, we do not have to consciously say, "OK, biceps, it's your turn to work. Hey, triceps, not so much there. Take it easy." Our neuromuscular system handles the chore all on its own—and usually does so with deceptive ease.

critical thinking questions

1. With respect to muscle function and cooperative action between muscles, define the following terms and give examples of their applications: *agonists, neutralization, stabilization, antagonists,* and *coactivation.*

2. Describe the muscle control formula outlined in this chapter. Apply the formula to any multijoint exercise (e.g., squat, lat pull-down).

3. Define movement efficiency versus movement economy. In addition, briefly describe the difference between mechanical and metabolic efficiency, and give specific sport examples of when one of these two forms of efficiency is most important. Hint: Think of the terms *mechanical* and *metabolic* as they refer to short-duration explosive or high-force events versus longer-duration more aerobic-type events, respectively.

4. Describe several types of actions or conditions that contribute to movement inefficiency, and give specific applications.

5. What are the benefits of performing kinematic, kinetic, and electromyographic assessments of human movement?

suggested readings

Basmajian, J.V., & DeLuca, C. (1985). *Muscles alive* (5th ed.). Baltimore: Williams & Wilkins.

Cappozzo, A., Marchetti, M., & Tosi, V. (Eds.). (1992). *Biolocomotion: A century of research using moving pictures.* Rome: Promograph.

Robertson, D.G.E., Caldwell, G.E., Hamill, J., Kamen, G., & Whittlesey, S.N. (2004). *Research methods in biomechanics.* Champaign, IL: Human Kinetics.

eight

Fundamentals of Posture, Balance, and Walking

objectives

After studying this chapter, you will be able to do the following:

- Describe movement concepts related to posture and balance

- Describe standing, sitting, and lying postures

- Explain the mechanisms of postural control

- Describe developmental aspects of balance and balance dysfunction

- Define walking and the gait cycle

- Describe the components of gait analysis

- Explain the role of lower-extremity muscles in the control of walking

- Describe gait development across the life span

- Describe examples of pathological gait

> There is no such thing as perpetual tranquility of mind while we live here; because life itself is but motion.
>
> —Thomas Hobbes (1588-1679)

ovement is essential for all life functions, from motion at the molecular level to whole-body actions. Within the body, movement of blood through the cardiovascular system and air through the respiratory system, for example, serves life-sustaining functions. Fundamental tasks such as eating require movement as well. On a larger scale, one of the most necessary things we do is move from one place to another, whether it is an infant's crawling, an adult's walking, or an athlete's sprinting or jumping.

Why move? We move because it is necessary for our survival. We move because it is essential for performing tasks of daily living. And we move because it allows us to experience the joy and exhilaration of dance, physical exercise, and competitive sports. As British author Laurence Sterne notes, "So much of motion, is so much of life, and so much of joy" (Sterne, 1980, p. 345).

We begin this chapter by exploring fundamental movement forms, beginning with posture and balance, followed by examination of walking. Our description of each movement form includes assessment of muscle control and takes a life-span approach that includes discussion of developmental aspects and how the movements differ in children, adults, and older individuals. Movement dysfunction due to injury, disease, or congenital factors is also considered.

Posture and Balance

From the first hesitant steps of a 1-year-old child to the remarkable feats of balance seen in circus performers and daredevils, the human body often must struggle to maintain its balance. To do this, people use information from several sensory systems to align their body segments. This positioning, or alignment, of the body is the essence of posture. In much the same way as stacking blocks crookedly makes for an unstable structure prone to falling, "crooked" body postures are also unstable. To produce a stable posture, the body's segments must be aligned correctly, just as the architect of a skyscraper must begin with a strong foundation and precisely align each successive story one upon another.

> Posture accompanies movement like a shadow.
>
> —Sir Charles Sherrington

Proper posture and balance are essential for the efficient performance of both simple daily tasks and more complex movement patterns. Improper posture and loss of balance can result in poor performance, injury, or even death.

Posture can be defined simply as the alignment, or position, of the body and its parts or more expansively as "a position or attitude of the body, the relative arrangement of body parts for a specific activity, or a characteristic manner of bearing one's body" (Smith, Weiss, & Lehmkuhl, 1996, p. 401). Clearly, these definitions are not restricted to a single alignment, or posture, but rather suggest an infinite number of possible postures. One might assume, for

In Others' Words: Charles Darwin

Great pain urges all animals, and has urged them during endless generations, to make the most violent and diversified efforts to escape from the cause of suffering. Even when a limb or other separate part of the body is hurt, we often see a tendency to shake it, as if to shake off the cause, though this may obviously be impossible.

Thus a habit of exerting with the utmost force all the muscles will have been established, whenever great suffering is experienced. As the muscles of the chest and vocal organs are habitually used, these will be particularly liable to be acted on, and loud, harsh screams or cries will be uttered. But the advantage derived from outcries has here probably come into play in an important manner; for the young of most animals, when in distress or danger, call loudly to their parents for aid, as do the members of the same community for mutual aid. (Darwin, 1998, p. 29)

—Charles Darwin, The Expression of the Emotions in Man and Animals

example, a standing, sitting, or lying posture or some other distinctive posture appropriate for a particular purpose or task.

To maintain a particular posture or change from one posture to another requires control of the body's alignment. This control of posture is termed balance. **Balance** can be defined as the maintenance of postural stability, or equilibrium, and often is used synonymously with the term **postural control**.

The concepts of posture and balance, although distinct from one another, are critically interdependent. Static postures often are difficult to maintain because the body and its parts continuously respond to forces (e.g., gravity, muscle) that tend to alter body alignment and disrupt the static equilibrium of the system. The body must continuously make subtle adjustments, usually through muscle actions, to maintain the equilibrium necessary for postural stability.

Functions of Posture and Balance

Posture and balance obviously are essential for all tasks we perform. Improper posture or loss of balance can negatively affect performance, decrease movement efficiency, and increase the risk of injury. Proper posture serves three functions: (1) maintenance of body segment alignment in any position: supine, prone, sitting, and standing; (2) anticipation of change to allow engagement in voluntary, goal-directed movements; and (3) reaction to unexpected disturbances, or perturbations, in balance (Cech & Martin, 2002).

Posture should not be considered a static phenomenon. Even while people stand or sit still, there are inevitable small fluctuations in their positions, or posture. Postural positions are maintained within ranges of movement. For a guard at Buckingham Palace, whose intent is to remain perfectly still, these fluctuations are imperceptible. But for someone waiting in line to purchase a movie ticket, the movements are considerably greater. Thus, posture from one moment to the next can involve varying amounts of movement, referred to as **postural sway**. Control of postural sway is maintained by muscle action.

Types of Posture

Most of us have been receiving postural advice since childhood. Parents admonish their children to "stand up straight" and "pull your shoulders back." Drill instructors scream at military recruits to "stand at attention." Teachers advise their students to "sit tall" and tell them "don't lean back in your chair." These admonitions try to persuade us to maintain good posture. They also suggest several different types of posture.

Static Postures

Many of our waking hours are spent either standing or sitting. Because these postures typically involve little movement, we refer to them as **static postures**. While resting or sleeping, we usually assume some type of static lying posture. It is important to emphasize that static postures are not completely motionless. Stationary, or static, postures typically involve slight movement, or swaying. Therefore, they sometimes are also referred to as **steady-state postures**.

STANDING POSTURE   The notion of a "normal posture" might erroneously be interpreted as meaning there is a single "best" posture. Given the variability in anatomical structure and physiological function (see chapter 1), no single posture is recommended for everyone. The normal posture for individuals depends on many factors, including their body type, joint structure and laxity, and muscular strength. Despite these inherent interindividual differences, certain characteristics are associated with good upright posture. In perpendicular standing, these characteristics include the following:

- Head is held in an erect position.
- Body weight is distributed evenly between the two feet.

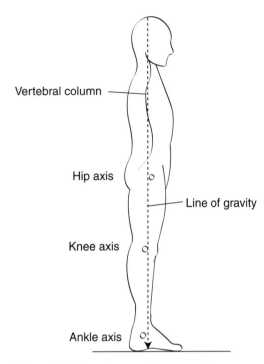

Vertebral column

Hip axis

Line of gravity

Knee axis

Ankle axis

Figure 8.1 Line of gravity during standing.

• From a frontal view, bilateral structures (e.g., iliac crests, acromion processes) are at the same horizontal level.

• From a sagittal plane (side) view, the line of gravity passes posterior to the cervical and lumbar vertebrae, anterior to the thoracic vertebrae, posterior to the hip joint, and anterior to the knee and ankle joints (figure 8.1 and table 8.1).

• Appropriate spinal curvatures are evident in the cervical, thoracic, and lumbar regions.

This upright posture is uncomfortable if held for an extended period. Most people who must stand for prolonged periods adopt various alternative, and more comfortable, postures (Smith, Weiss, & Lehmkuhl, 1996). One alternative posture is asymmetric standing, characterized by a weight shift to one leg that is fully extended. By fully extending the knee, the line of gravity passes anterior to the knee joint, creating an extensor moment and reducing the need for quadriceps muscle activity (figure 8.2). Another alternative posture adopts a wide base of support with both legs fully extended and arms held behind the back or crossed on the chest. A third alternative is the **nilotic stance**, in which the individual stands on one leg with the opposite leg used to brace the standing knee (i.e., like a flamingo).

Table 8.1   Normal Alignment in the Sagittal Plane

Joint	Line of gravity	Gravitational moment	Opposing forces	
			Passive opposing forces	Active opposing forces
Atlantooccipital	Anterior	Flexion	1. Ligamentum nuchae	Posterior neck muscles
	Anterior to transverse axis for flexion and extension		2. Tectorial membrane	
Cervical	Posterior	Extension	1. Anterior longitudinal ligament	
Thoracic	Anterior	Flexion	1. Posterior longitudinal ligament	Extensors
			2. Ligamentum flavum	
			3. Supraspinous ligament	
Lumbar	Posterior	Extension	1. Anterior longitudinal ligament	
Sacroiliac	Anterior	Flexion-type motion	1. Sacrotuberous ligament	
			2. Sacrospinous ligament	

Joint	Line of gravity	Gravitational moment	Opposing forces	
			Passive opposing forces	Active opposing forces
			3. Sacroiliac ligament	
Hip	Posterior	Extension	1. Iliofemoral ligament	Iliopsoas
Knee	Anterior	Extension	1. Posterior joint capsule	
Ankle	Anterior	Dorsiflexion		Soleus

Adapted from Norkin & Levangie 1992.

SITTING POSTURE   Many people spend long hours in a sitting, or seated, position, whether at home, work, or school. Correct sitting postures can reduce spinal loading and the chance of injury. Incorrect postures, in contrast, can increase the risk of injury. Historically, an ideal sitting posture has been characterized by the ischial tuberosities acting as the major base of support, anterior pelvic tilt (which maintains appropriate lumbar curvature), spinal support provided by a slightly inclined seatback, and the feet contacting the floor to share in supporting the body's weight (figure 8.3a).

Poor sitting posture, usually characterized by a slouched position (figure 8.3b), results in posterior pelvic tilt (which flexes the lumbar spine and reduces the lumbar curvature), increased stretching and weakening of the posterior annulus fibrosis, and increased flexor torque created by an anterior shift in the line of gravity (Neumann, 2002). All these

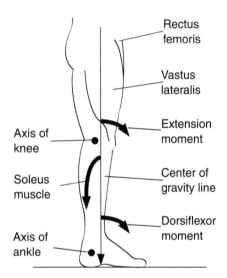

Figure 8.2   Effect of fully extending knee on quadriceps activation levels.

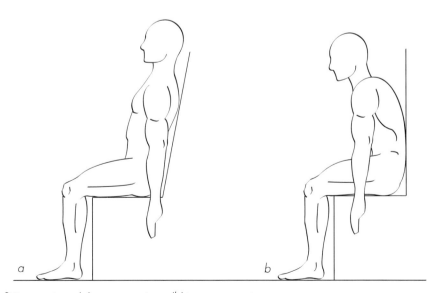

Figure 8.3   Sitting posture: (a) proper posture, (b) improper posture.

characteristics may increase the chance of lumbar disc injury and low back pain. A slouched posture also affects more than just the lumbar region. Increased lumbar flexion contributes to excessive flexion (kyphosis) in the thoracic spine and a protracted, or thrust-forward, head (see figure 8.3b). This head position puts added stress on the vertebrae, muscles, and ligaments of the neck and shoulders.

Well-designed chairs can facilitate proper sitting postures, while poorly designed chairs make good postures difficult and may contribute to musculoskeletal disorders such as intervertebral disc degeneration, low back pain, inflexibility, and loss of joint range of motion.

The notion that an "ideal" sitting posture exists recently has been challenged. "Despite the myth perpetuated in many ergonomic guidelines regarding a single 'ideal' posture for sitting, the ideal sitting posture is one that continually changes, thus preventing any single tissue from accumulating too much strain" (McGill, 2002, p. 104). McGill convincingly argues that one of the greatest risks for low back pain is the prolonged nature of sitting, which increases the risk of disc herniation. Further, low back health is facilitated by properly using an ergonomic chair to facilitate frequent *changes* in sitting posture, periodically getting out of the chair and assuming a relaxed standing position, and performing an exercise routine at some time during the workday, preferably *not* in the early morning when the back is more susceptible to injury (McGill, 2002).

LYING POSTURE   When resting or sleeping, we normally assume a **recumbent**, or lying, posture because it is the least physiologically demanding. This posture rotates the action of gravity from its longitudinal orientation in standing (figure 8.4a) and sitting to a transverse direction relative to the long axis of the body. Basic lying postures include lying face down (**prone**), lying face up (**supine**), and lying on one side or the other (figure 8.4b). Each of these lying postures has advantages and disadvantages.

Surface characteristics play an important role in spinal alignment and how forces are applied to the body. On a very hard surface, only specific body areas (e.g., hip, shoulder) come in contact with the surface (figure 8.5a). This creates localized pressure points that can be uncomfortable or even injurious. Sleeping on too soft a surface also can cause problems, including excessive lumbar flexion when in a supine position, exaggerated lumbar extension in a prone posture,

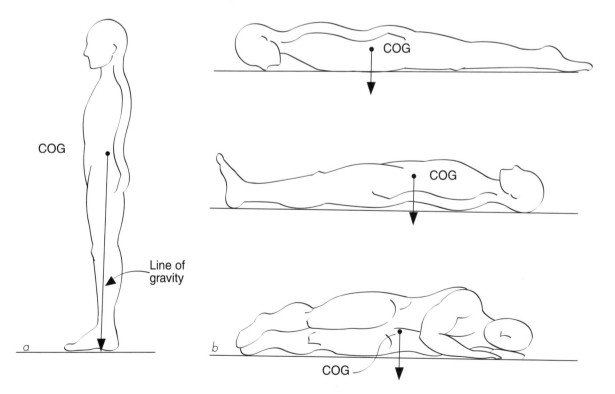

Figure 8.4 Line of gravity standing versus lying: *(a) standing*; *(b)* prone, supine, and side postures.

and lateral spinal curvatures when side lying (figure 8.5b). The surface and head support (e.g., pillow) should be consistent with maintaining proper spinal alignment (figure 8.5c).

Dynamic Posture

Dynamic posture is the posture of motion, as seen in walking, running, jumping, throwing, and kicking. Each of these movements requires continual changes in the position of the trunk and extremities that must be controlled to maintain the dynamic equilibrium necessary for task completion. Loss of dynamic postural control can result in task failure (e.g., stumbling or falling) and possible injury.

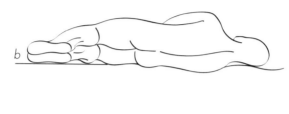

Types of Postural Control

Balance, or postural control, is necessary in dynamic situations and for maintaining static (stationary) positions. There are four types of postural control: static, reactive, anticipatory, and adaptive (Cech & Martin, 2002).

Figure 8.5 Lying postures and surfaces: (a) hard surface, (b) soft surface, (c) proper alignment with pillow and surface.

Static postural control involves strategies for keeping the vertical projection of the body's center of gravity (i.e., line of gravity) within the base of support (BOS). In normal upright standing, the body commonly sways slightly from side to side and forward and backward. When the body sways in one direction, the body's line of gravity moves slightly away from the center of the BOS toward its edge. Arresting this movement and moving the line of gravity back toward the center requires muscle action. For example, if the body sways anteriorly (forward), creating a gravitational dorsiflexion moment (torque) at the ankle, the soleus must be recruited to create a plantar flexion coun-

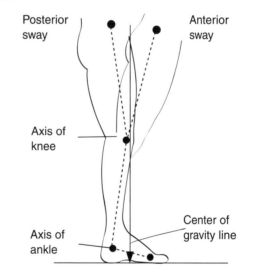

Figure 8.6 Postural sway: anterior sway and posterior sway.

termoment (see chapter 6) to recentralize the line of gravity (figure 8.6). Similarly, a posterior (backward) body lean creates a gravitational plantar flexion moment at the ankle. The tibialis anterior must then be recruited to generate a dorsiflexion countermoment to bring the line of gravity back toward the center of the BOS and reestablish postural equilibrium .

When unexpected events (e.g., slipping or tripping) cause the line of gravity to move away from the center of the base of support, **reactive postural control** is necessary to maintain balance and prevent a fall. If the line of gravity moves away from the BOS center but does not exceed the BOS, the neuromuscular system quickly recruits muscles to regain postural stability, through what is termed a righting response. In contrast, if the line of gravity moves outside the base of support, the body must employ a different strategy to maintain balance. This strategy typically involves a change in the BOS, such as when a person trips, falls forward, and takes a step to prevent a fall. By taking this step to create a new and larger BOS, the person is exercising reactive postural control.

Anticipatory postural control involves making movements in anticipation of events that will cause postural changes and potential loss of balance. For example, before picking up a heavy suitcase with the right hand, the carrier leans her trunk to the contralateral (left) side to counteract the anticipated load of the suitcase; a rider standing in the aisle of a stationary bus leans forward or grabs the rails in anticipation of the bus's acceleration; a football player about to collide with an opponent leans in the direction of the impact. In all these examples, the individual makes anticipatory movements in advance of predicted disruptive events.

Adaptive postural control allows us to modify our movements in response to situational changes. Postural control depends on environmental conditions and task demands and thus is context dependent (Enoka, 1994). Control of posture is not limited to postural reflexes. "[T]he control of posture is not simply based on a set of reflex responses, nor is it a preprogrammed response that is triggered by a disturbance. Rather, the control of posture is an adaptable feature of the motor system that relies on the integration of afferent input and efferent output" (Enoka, 1994, p. 250). Postural responses can adapt to repeated disturbances (perturbations), allowing us to alter our responses and effectively learn how to better respond to the postural disturbance.

Mechanisms of Postural Control

Before the neuromuscular system can initiate corrective postural actions, it needs to receive information from the visual, vestibular, and somatosensory systems. This information, provided by the nervous system, communicates the nature of the physiological, mechanical, and informational perturbations affecting the body and allows the nervous system to recruit the appropriate muscles required to reestablish equilibrium (figure 8.7).

The postural sway inherent to standing is detected by sensory receptors in the eyes (visual), ears (vestibular), skin (tactile), and joints (proprioceptive). Based on the information supplied by these receptors, the nervous system responds by recruiting muscles to create moments (torques) that counteract the gravitational moments created by the line of gravity sway away from the center of the base of support.

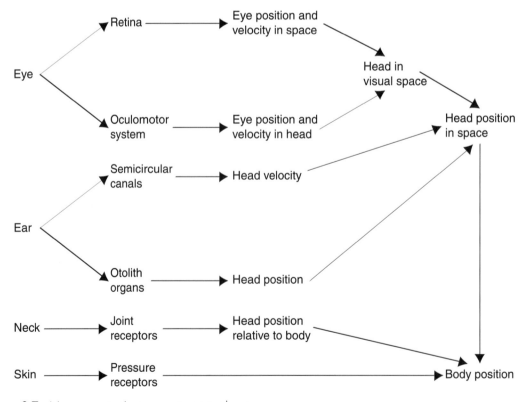

Figure 8.7 Nervous system's response to postural sway.

Reprinted from R.H.S. Carpenter, 1984, *Neurophysiology* (New York: Edward Arnold).

Similarly, in a dynamic task, a gymnast attempting to walk, run, and jump along a narrow balance beam needs continuous sensory feedback to maintain balance and proceed along the beam. Compromised information from any of the sensory systems may result in loss of balance and a fall from the beam.

When someone stumbles or trips while walking, these same sensory systems rapidly provide feedback to the neuromuscular system, which then attempts to recruit the muscles needed to recover balance (e.g., by making a corrective step or reaching for a rail or table for support). Incorrect, incomplete, or delayed sensory information may lead to an unsuccessful recovery and a resulting fall.

Postural Alterations and Perturbations

Deviations from normal postures can be caused by physiological, psychological, environmental, and anatomical factors. A marathoner experiencing physiological fatigue may alter her running posture to compensate for altered muscle activation patterns and joint mechanics; a worker experiencing depression may walk with a lethargic, slumped posture; and an older adult navigating a slippery staircase might adopt a different stepping posture than would a younger person. Anatomically, abnormal posture may be caused by structural defects. For example, a person with a leg-length discrepancy (i.e., one leg longer than the other) may make compensatory postural adjustments, such as a lateral pelvic tilt, that alters joint mechanics.

Altered posture can also be caused by task-dependent disturbances of motion, or **perturbations**. The posture a warehouse worker uses to pick up a dropped pencil, for example, likely differs from the posture he adopts when picking up a heavy box.

Researchers have used several paradigms to study how people react to perturbations in the environment. One of the most common approaches uses a moving platform to induce horizontal perturbations to standing posture. Investigators then can measure response times, muscle activity patterns, segmental kinematics, and joint forces and torques to assess the mechanisms of postural control. The platforms can be moved in either the anterior–posterior (forward–backward) or lateral (side-to-side) directions, with some also able to change the velocity and inclination of the platform.

The postural responses to the moving platform are task specific and vary with the size of the supporting surface, direction of motion, location and magnitude of the applied force, initial posture at the moment of perturbation, and velocity of perturbation (Levangie & Norkin, 2001).

When the platform moves in the anterior–posterior direction, for example, the subject can adopt joint-specific strategies in response to the perturbation. An *ankle strategy* involves activation of either the plantar flexors (e.g., soleus) or the dorsiflexors (e.g., tibialis anterior) to control ankle motion. If the stationary platform suddenly moves anteriorly (forward), the body tends to fall backward. This creates movement-induced plantar flexion at the ankle. To compensate, the neuromuscular system recruits dorsiflexors to pull the lower leg forward toward the vertical position. If the platform moves posteriorly (backward), the body falls forward, creating movement-induced dorsiflexion at the ankle. In response, plantar flexor muscles are recruited to pull the lower leg backward toward vertical.

The subject could also employ a *hip strategy*. Sudden forward movement of the platform would cause hip extension. Hip flexors (e.g., iliopsoas) would be recruited to maintain an upright posture. Backward platform movement would induce hip flexion, with hip extensors (e.g., gluteus maximus) then required to vertically realign the trunk. If these strategies prove insufficient to reestablish balance, the subject could employ a *stepping strategy* to maintain stability by increasing the base of support.

Developmental Considerations

At both ends of the life span, postural control is less effective than during healthy adulthood. In the early years, infants are still developing their sensorimotor systems and gaining the strength and coordination needed to adopt a bipedal, upright posture. In older adults, disease, strength loss, compromised sensory function, and general motor decline can negatively affect balance and increase the risk of injury.

Posture and Balance in Infants

Postural development in an infant begins with control of the head and proceeds to the trunk. Automatic postural reactions, summarized in table 8.2, develop during the first year of life. Also within the first year of life, an infant reaches postural milestones of sitting and standing and begins developing the balance strategies required for postural control.

Clearly during this first year, infants respond to environmental cues in what is termed **perception–action coupling**. An infant perceives optical flow (i.e., visual image changes due to motion) and coordinates this information to generate postural responses. Infants scale their postural responses to the visual information they receive.

In the first few years of development, an infant reaches numerous posturally related **motor milestones**. According to the Bayley Scales of Infant Development, these include holding the head erect and steady (average age: 1.6 months), sitting with slight support (2.3 months), sitting alone momentarily (5.3 months), rolling from back to front (6.4 months), sitting alone steadily (6.6 months), pulling to stand (8.1 months), standing up by furniture (8.6 months), stepping movements (8.8 months), standing alone (11.0 months), walking alone (11.7 months), walking backward (14.6 months), walking up stairs with help (16.1 months), and jumping off the floor with both feet (23.4 months). The order of these milestones is relatively consistent, with each skill building on earlier ones, but there is a wide age range at which different children reach each milestone. In addition, other scales (e.g., Shirley Sequence) may list somewhat different average ages for each milestone (Haywood & Getchell, 2001).

Balance continually improves through infancy and childhood, but the exact course of the improvement depends on the task. Perception–action coupling must be developed and refined for each new skill, and although the general trend moves toward improved balance, plateaus or even decrements in balance performance may occur along the way (Haywood & Getchell, 2001).

Posture and Balance in Older Adults

Balance impairment in older adults is all too common and a major contributor to the high incidence of falls in the elderly. Falls are the leading cause of fractures in older adults. Fall-related

Table 8.2   Postural Reactions in Infants

Reaction	Starting position	Stimulus	Response	Time
Derotative righting	Supine	Turn legs and pelvis to other side	Trunk and head follow rotation	From 4 mo
	Supine	Turn head sideways	Body follows head in rotation	From 4 mo
Labyrinthine righting reflex	Supported upright	Tilt infant	Head moves to stay upright	2-12 mo
Pull-up	Sitting upright held by 1 or 2 hands	Tip infant backward or forward	Arms flex	3-12 mo
Parachute	Held upright	Lower infant toward ground rapidly	Legs extend	From 4 mo
	Held upright	Tilt forward	Arms extend	From 7 mo
	Held upright	Tilt sideways	Arms extend	From 6 mo
	Held upright	Tilt backward	Arms extend	From 9 mo

Reprinted, by permission, from K.M. Haywood and N. Getchell, 2001, *Life span motor development*, 3rd edition (Champaign, IL: Human Kinetics), 90.

fractures most commonly occur in the radius (from falling on an outstretched arm) and femur (from impact loading at the hip). Hip fractures number more than 300,000 in the United States annually and are a major cause of disability and death.

Many age-related changes compromise postural control ability, including the following:

- Decline in visual skills (e.g., less able to discern contours and depth cues)
- Visual pathology (e.g., cataracts, macular degeneration)
- Compromised vestibular function (e.g., loss of vestibular hair cells, reduction of nerve fibers in cranial nerve VIII)
- Increased postural sway
- Difficulty using built-in redundancy of the sensory systems
- Slowing of central control mechanisms
- Conflicting sensory inputs
- Slowing of reaction and response times
- Increased reflex latency
- Declines in strength
- Changes in muscle fiber type
- Attention deficits
- Increased sensitivity to absence of visual input
- Slower anticipatory responses
- Diminished control of line of gravity and base of support
- Impaired ability to control compensatory stepping movement
- Psychological state
- Pain
- Decreased joint range of motion
- Decreased balance time and joint torque production
- Anatomical and structural changes (e.g., loss of fluid from intervertebral discs, disc degeneration)
- Increase in "flexed," or "stooped," posture

Summarized from Cech & Martin, 2002; Maki & McIlroy, 1996; Levangie & Norkin, 2001.

Each of these factors contributes to declines in balance, and their composite effect determines the severity of overall decline for a given individual. Some of the changes are inevitable, while others can be ameliorated if interventions are begun soon enough. The first step in arresting the decline or even improving balance in older adults includes awareness and identification of the factors specific to a particular individual. Controllable factors include muscular weakness and imbalance, joint range of motion, psychological state, and control of compensatory movements. The controllable factors can be improved through a combination of education, biofeedback, postural retraining, therapeutic exercise, orthotics, strength training, practice of movement forms (e.g., tai chi) known to enhance postural awareness and balance, and practice of recovery from perturbations.

Postural Dysfunction

Many factors contribute to poor posture and postural dysfunction, including pain, decreased joint range of motion, inflexibility, muscle weakness and imbalances, altered joint biomechanics, joint hypermobility and ligament laxity, altered sensation and proprioception, psychological state, adaptations to the environment, persistent adoption of poor posture (habituation), pregnancy, anatomical defects, fatigue, disease (e.g., muscular dystrophy, osteoporosis), and injury

(Neumann, 2002; Trew & Everett, 2001). Poor posture places abnormal loads on anatomical structures and increases the risk of musculoskeletal injury. Physiological functions such as respiration and circulation also can be negatively affected.

As described earlier, the spine's normal curvatures help the body accept compressive loads. Injury, disease, and congenital predisposition can cause deformities of the spinal column (e.g., abnormal structural alignment or alteration of spinal curvatures). These deformities often result in altered force distribution patterns and pathological tissue adaptations that may lead to or exacerbate other musculoskeletal injuries. As described in chapter 3, there are three primary spinal deformities: scoliosis, kyphosis, and lordosis. Classified by their magnitude, location, direction, and cause, these deformities can occur in isolation or in combination.

Walking

Of all the ways of "going forward," walking is by far the most common. Virtually everyone has experience walking and usually gives little thought to this mode of getting from one place to another. Although walking displays certain common characteristics, each of us develops a walking style uniquely our own. Most of us have experienced seeing the silhouette of someone in the distance, and though we cannot recognize him by facial features, we "know who it is by the way he walks." What is it about walking that allows us to identify a person at a distance based solely on how he walks? This largely involves the mechanics of walking style.

> There are many ways of going forward, but only one way of standing still.
>
> —Franklin D. Roosevelt (1882-1945)

Walking undoubtedly has been the subject of more research than any other movement form. Several excellent books (some listed in the suggested readings at the end of this chapter) are devoted solely or primarily to exploring the intricacies of human walking. Thousands upon thousands of research articles published in the last century have dealt with various aspects of walking, as have many notable books devoted to the topic. As a result, we arguably know more about walking than any other human movement form. Given the volume of information on walking, clearly we can only scratch the surface in the limited space afforded here. We provide the basics and hope that the interested reader will pursue further study of walking using the many comprehensive resources available.

Terminology

Several terms are important to our understanding of basic movement forms. **Locomotion** is defined as "an act or the power of moving from place to place" (Mish, 1997, p. 684). This broad definition obviously includes common movement forms such as walking and running but also embraces less common movement modes such as skipping, crawling, sliding, swimming, and even playing leapfrog or doing cartwheels. All of these movements, and many others, are forms of locomotion because they fulfill the basic requirement of locomotion—getting from one place to another.

Gait refers to a particular form of locomotion. Most commonly, gait is used in the context of describing *walking gait* or *running gait*. (Note: Some references use gait only in reference to walking.)

Walking in humans can be defined as a form of upright, bipedal locomotion, or gait, in which at least one foot is always in contact with the ground. It is a cyclic activity involving the alternating action of the legs to advance the body forward. At first glance, walking might appear to be a simple movement task. After all, it's something we do every day with little conscious thought or consideration. Voluminous research over recent decades, however, has proved much the opposite; walking entails a complex set of neuromechanical events. The challenges of walking are even more profound for infants, older adults, and individuals affected by disease, injury, or congenital defect.

To study walking gait, we need to first understand several terms commonly found in the gait literature. As a person walks, each leg alternates between periods when the foot is in contact with the ground and when it is moving forward through the air (with no foot–ground contact). The period during which the foot contacts the ground is termed the **stance phase**, or stance. When the foot is not in ground contact, the leg is moving forward in what is termed the **swing phase**, or swing. Stance and swing describe the phasing of each leg independently of the other.

The stance phase for a given leg begins when that leg's foot first hits the walking surface. This first contact is called **initial contact**. (Note: Several other terms are found in the literature to describe this first contact. Among these are *heel contact, heel strike, foot contact*, and *foot strike. Initial contact* is preferred here, since it provides the most general description of the event. The other terms may be misleading. *Heel contact*, for example, suggests that the heel is the part of the foot that first contacts the ground. Although usually true, in some conditions such as paralysis or cerebral palsy the heel may not be the first contact point on the foot.) The stance phase continues until the foot leaves the ground in an event termed **toe-off**, or **takeoff**. Toe-off initiates the swing phase, which continues as the leg swings forward and the body advances until the next ipsilateral (i.e., same side) initial contact.

A **gait cycle** refers to the sequential occurrence of a stance and swing phase for a single limb. This period from initial contact of one leg until the next initial contact of the same (ipsilateral) leg is termed a **stride**. Each stride is made up of two steps, with a **step** defined as the period from initial contact of one leg to initial contact of the opposite (contralateral) limb. Strides and steps are measured by length and width as shown in figure 8.8. (Note: The term *stride*, as described here in its common clinical and biomechanical usage, is used differently in the track and field literature. In track and field circles, stride refers to the period from ipsilateral contact to contralateral contact, or what we describe here as a step.)

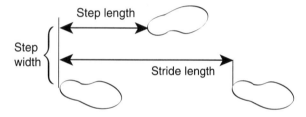

Figure 8.8   Step length, stride length, and step width.

In walking there obviously are periods when only one foot is in contact with the ground, when one leg is in its swing phase and the contralateral leg is solely supporting the entire body. This period is termed **single support**. When both feet are touching the walking surface at the same time (i.e., both legs are in their respective stance phases), the period is termed **double support**.

Another important term is **cadence**, defined as the step rate as measured by steps per minute. A person walking with a cadence of 120 steps per minute, for example, takes 120 steps (60 strides) every 60 seconds, so each step would take 0.5 seconds.

Walking speed (velocity) is determined by the mathematical product of cadence and step length. If a person has an average step length of 0.7 m and a cadence of 114 steps per minute, her walking speed is 79.8 m/min (0.7 m × 114 steps/min). Obviously, walking speed differs depending on a number of factors, including the person's age, size, and physical condition, as well as the purpose of the walk. A person who is late for an appointment, for example, will choose a faster walking speed than if she is taking a leisurely stroll in the park. Every individual has a free, self-selected, and comfortable walking speed that minimizes energy expenditure.

Gait Cycle

The gait cycle, as defined earlier, encompasses the stance–swing period of a single leg. The gait cycle is divided into phases, or subcycles. No universally accepted system exists for specifying these phases. We present two widely used systems that, although different, share many common elements. The first, a traditional system, identifies critical points in the stance phase (heel contact, foot flat, midstance, heel-off, and toe-off) and divides the swing phase into periods of initial swing (acceleration), midswing, and terminal swing (deceleration) (Levangie & Norkin, 2001) (figure 8.9).

Another widely adopted system is that proposed by Perry (1992), which divides the gait cycle according to periods (stance and swing), tasks (weight acceptance and single limb support during stance; limb advancement during swing), and phases (initial contact, loading response, midstance, terminal stance, preswing, initial swing, midswing, and terminal swing) (figure 8.10). (Note: A detailed comparison of these two systems is provided by Levangie & Norkin, 2001, pp. 438-443.) Other systems use similar terminology that differs only in some of the details in describing the gait cycle.

Gait Analysis

Gait analysis can range from simple measures of timing (**temporal analysis**) to mechanical analysis using sophisticated instrumentation and computer models. The goal of gait analysis is to provide scientific and clinical insights into both normal and pathological walking. Most if not all of the variables explored in gait analysis change with walking speed. The characteristics of a slow, leisurely walk differ greatly from those typical of a fast, determined walking gait. Although observational gait analysis has been performed for many centuries, scientific measurement has been limited to the last 100 years or so.

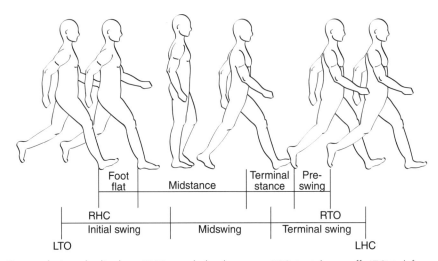

Figure 8.9 Gait cycle (standard) where RHC is right heel contact, RTO is right toe-off, LTO is left toe-off, and LHC is left heel contact.

Adapted from *Kinesiology of the Musculoskeletal System: Foundations for Physical Rehabilitation*, 2nd edition, D.A. Neumann, pg. 530, Copyright 2002, with permission from Elsevier.

Divisions of the Gait Cycle

Stride (Gait cycle)

Periods

Stance — Swing

Tasks

Weight acceptance — Single limb support — Limb advancement

Phases

Initial contact — Loading response — Midstance — Terminal stance — Preswing — Initial swing — Midswing — Terminal swing

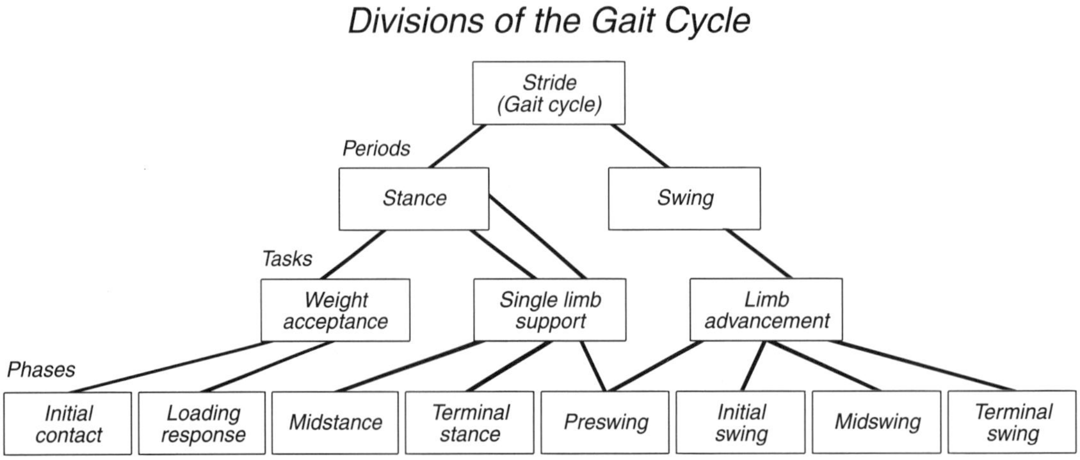

Figure 8.10 Gait cycle (Perry).

Adapted from *Kinesiology of the Musculoskeletal System: Foundations for Physical Rehabilitation*, 2nd edition, D.A. Neumann, pg. 532, Copyright 2002, with permission from Elsevier.

Temporal and Spatial Characteristics

Several timing (temporal) measures of gait are commonly used. Cadence, or step rate, is perhaps the simplest of these measures. The average free walking (i.e., self-selected speed) cadence for adults is about 113 steps per minute. Women typically walk with a higher cadence (117 steps/min) than do men (111 steps/min) to partially compensate for their shorter step length (Perry, 1992).

In temporally symmetrical gait (i.e., equal timing between sides), **step time** is simply the inverse of cadence. A cadence of 114 steps per minute, for example, represents a step time of 0.0088 minutes, or 0.53 seconds. Because two steps make up one stride, the stride time is 1.06 seconds.

In free walking, a person spends about 60% of the gait cycle in stance and 40% in swing (figure 8.11). As walking speed increases, the ratio of stance to swing approaches 50:50. To enhance stability, older adults spend more time in stance and less in swing.

As described earlier, single support is the phase when only one foot contacts the ground; in double support, both feet are in ground contact. Each gait cycle includes two periods of single support and two periods of double support. The relative duration of each of these periods is shown in figure 8.12.

Three common spatial gait measures are step length, stride length, and step width

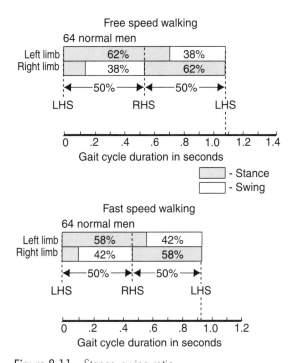

Figure 8.11   Stance–swing ratio.

Adapted, by permission, from M.P. Murray, D.R. Gore, and B.H. Clarkson, 1971, "Walking patterns of patients with unilateral hip pain due to osteoarthritis and avascular necrosis," *The Journal of Bone and Joint Surgery* 53(2): 259-274.

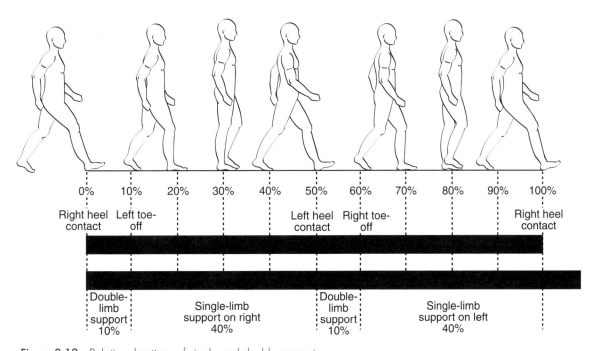

Figure 8.12   Relative durations of single and double support.

Reprinted, by permission, from J. Perry, 1992, *Gait analysis—normal and pathological function* (Thorofare, NJ: Slack, Inc.), 10.

(see figure 8.8). Because of their taller stature, men on average have greater (~14%) step length and stride length than women. Men and women average a stride length of 1.46 m and 1.28 m, respectively (Perry, 1992). Step width averages 7 to 9 cm. In normal gait, the foot also angles outward (laterally) by about 7° from the direction of progression.

Combining temporal (time) and spatial (distance) measures tells us how fast a person is walking. The average adult walks at a speed of about 3 mph (1.34 m/s), with women self-selecting a slightly·slower speed than men (Neumann, 2002). Walking speed can vary considerably based on a number of factors including age, gender, physical condition, environmental conditions, and purpose. (Note: Some sources refer to walking speed, while others describe walking velocity. As described in chapter 6, speed is a scalar measure of how fast a person is moving, while velocity is a vector measure indicating both how fast and in what direction. The difference between speed and velocity in the context of describing gait is usually inconsequential because the direction is forward. Hence, the terms velocity and speed often are used interchangeably to describe how fast a person walks.)

Gait velocity is a function of stride length and stride rate according to the following equation:

$$\text{velocity} = \text{stride length} \times \text{stride rate} \quad (8.1)$$

or equivalently (for symmetrical gait)

$$\text{velocity} = \text{step length} \times \text{cadence} \quad (8.2)$$

For example, using equation 8.2, a person with a cadence of 112 steps per minute and a step length of 0.70 m would have a walking velocity of 78.4 m/min (1.31 m/s or nearly 3 mph).

Muscle Activity and Control of Walking

 Proper magnitude and timing in movement of the lower-extremity musculature are essential for normal walking gait. Even subtle muscle deficiencies can markedly affect gait performance. Significantly compromised muscle function, as in paralysis, may render normal gait impossible. In older adults and individuals with injury, pain is often a primary factor in gait alteration. Consider, for example, times you have been injured (e.g., sprained ankle). The pain from the injury likely caused you to alter your muscle activation patterns and gait mechanics. You probably tired more quickly than normal because altered gait is physiologically less efficient than self-selected normal gait, and you may also have felt discomfort or pain in other body regions (e.g., contralateral knee or hip) resulting from the mechanical changes in how you walked.

In normal gait, muscles of the hip, knee, and ankle are primarily responsible for controlling joint motion. Excellent detailed analysis of muscle action during gait can be found in a number of sources (e.g., Levangie & Norkin, 2001; Neumann, 2002; Perry, 1992; Rose & Gamble, 1994).

Highlights of muscle activity during phases of the gait cycle (using the system described by Perry, 1992) include the following:

• *Initial contact and loading response:* Hip abductors (upper fibers of the gluteus maximus, gluteus medius, gluteus minimus, tensor fascia lata) eccentrically prevent excessive lateral pelvic tilt; knee extensors (quadriceps: vastus intermedius, vastus lateralis, vastus medialis, rectus femoris) eccentrically prevent excessive knee flexion; ankle dorsiflexors (tibialis anterior, extensor hallucis longus, extensor digitorum longus, peroneus tertius) eccentrically control ankle plantar flexion. These actions work to accomplish weight acceptance, pelvic stabilization, and deceleration of the body.

• *Midstance:* Plantar flexors (soleus, gastrocnemius, tibialis posterior, flexor digitorum longus, flexor hallucis longus, peroneus longus, peroneus brevis) eccentrically control tibial advancement over the foot. The gastrocnemius (with two-joint action at the ankle and knee) also helps stabilize the knee.

• *Terminal stance:* Plantar flexors act concentrically to assist with push-off.

- *Preswing and initial swing:* Hip flexors (iliacus, psoas, rectus femoris, gracilis, sartorius) initiate hip flexion early in the swing phase.

- *Initial swing and midswing:* Ankle dorsiflexors (tibialis anterior, extensor hallucis longus, extensor digitorum longus, peroneus tertius) act concentrically to dorsiflex the foot at the ankle to guarantee toe clearance.

- *Late midswing and terminal swing:* Ankle dorsiflexors eccentrically control ankle plantar flexion; hamstrings (semitendinosus, semimembranosus, long and short heads of biceps femoris) eccentrically control knee extension in preparation for initial contact.

Any disturbance in the sequencing or level of muscle activation may alter gait mechanics and create the need for compensatory muscle action. Several examples of altered, or pathological, gait are described later in the chapter.

Although we have focused on the lower-extremity muscles, the arms and trunk also play an essential role in gait mechanics. The arms, for example, typically work in alternating fashion with the legs to provide a counterbalancing effect. When walking, the left arm moves synchronously with the right leg, while the right arm and left leg move in concert with one another. If you doubt the importance of the arms in gait, try walking briskly or running with your hands straight at your sides or placed on top of your head. You'll instantly feel the discomfort and uncoordinated motion this change in normal mechanics causes.

Life-Span Perspective

Newborn infants are essentially immobile. Developmental changes during the first year allow infants to take their first halting steps, and by the ages of 3 to 4 years, children usually demonstrate a mature walking gait pattern. The ability to alter walking speed, change pace and direction, and react to hazards continues to develop into adulthood and is maintained well into one's 50s and beyond. Numerous age-related changes make individuals in the 60-plus age range more susceptible to declines in gait function.

Infants and Children

The road from infant immobility to upright, bipedal walking gait is marked by a number of motor milestones. To complete these progressively more demanding tasks, an infant must develop the needed strength and neuromuscular coordination, and the environment must provide a flat, firm surface with sufficient friction to allow locomotion. The first forms of infant locomotion typically are crawling and creeping. **Crawling** is progressing on hands and stomach; **creeping** is moving on hands and knees. The normal developmental progression is (1) crawling with the chest and stomach on the floor, (2) symmetrical leg movements that create low creeping with the stomach off the floor, (3) back and forth rocking in a high creep position, and (4) creeping with alternating action of the arms and legs (Haywood & Getchell, 2001).

Before the end of their first year, most infants have developed sufficiently to stand without support and take their first steps. By 2 years of age, most of the characteristics of mature walking are in place. Over the next several years, children refine these characteristics through increased strength, coordination, and joint range of motion. By about 5 years, children exhibit a mature gait pattern that shows most, if not all, of the characteristics of proficient walking, which include the following:

- Absolute increase in stride length (resulting from increased leg length and greater force application and leg extension at push-off)
- Change from flat-foot planting to heel-to-forefoot pattern, resulting in increased range of motion
- Reduced out-toeing
- Narrowed base of support laterally to emphasize anterior–posterior force application
- Walking pattern that includes full knee extension at initial contact (heel strike), followed by slight knee flexion through midstance and full extension again during push-off

- Pelvis rotation to allow full range of leg motion and reciprocal movement of the upper and lower body segments

- Improved balance and reduction in forward trunk inclination

- Coordinated arm and leg movements so that the opposite arm and leg move forward and backward in unison

Adapted from K.M Haywood, and N. Getchell. (2001). *Life span motor development* (Champaign, IL: Human Kinetics), 122-123.

From this point on, only subtle gait changes are seen through the rest of childhood and into the adult years.

Older Adults

Older adults show a number of predictable changes in their walking gait. Although some older adults maintain gait patterns similar to younger adults well into their 60s and 70s, most individuals in this age range begin to exhibit changes that include slower walking speed; shorter step length and relative swing time; longer relative stance time; decreased maximum walking speed, joint range of motion, and cadence; and increased base of support, double limb support, and use of visual scanning (Bohannon, 1997; Judge, Ounpuu, & Davis, 1996; Ostrosky, VanSwearingen, Burdett, & Gee, 1994). Chronic conditions such as heart disease, arthritis, osteoporosis, or pain may exacerbate these changes or lead to their earlier onset.

The consequences of age-related changes in gait, coupled with declines in reaction time, strength, endurance, and visual acuity, greatly increase the chance of accidents such as tripping, slipping, and falling. Injuries associated with falls in older adults are a serious and growing public health problem. Considerable research is directed at characterizing the mechanics of falls and devising strategies to reduce their incidence. The human toll and monetary costs (in the billions of dollars) associated with the more than 300,000 annual fall-related hip fractures in the United States are but one example of how gait changes in older adults are integrally related to a major public health issue.

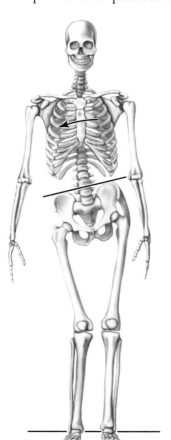

Figure 8.13 Trendelenberg gait. Note the ipsilateral pelvic drop.

Pathological Gait

Countless conditions can contribute to dysfunctional, or pathological, gait including disease, injury, paralysis, anatomical abnormalities, and congenital defects. Comprehensive examination of pathological gait is beyond the scope of this book, but we present an overview of three examples that demonstrate how anatomical, physiological, and environmental conditions can alter the gait pattern, increase the loads placed on anatomical structures, stress physiological systems, and exaggerate the chance of injury.

The first example, **Trendelenberg gait**, results from weakness or paralysis of the hip abductors (e.g., gluteus medius, gluteus minimus). In normal walking, during early to midstance of a given leg, the pelvis tends to drop to the contralateral side. This drop is controlled by eccentric action of the hip abductors. Compromised hip abductor action allows for the excessive pelvic drop characteristic of Trendelenberg gait. To compensate for this pelvic drop, subjects typically lean their trunks toward the side of the support leg (figure 8.13).

The second example involves gait changes characteristic of people with anterior cruciate ligament (ACL) deficiencies. The ACL ligament in the knee restricts anterior translation of the tibia relative to a fixed femur or, conversely, limits posterior movement of the femur relative to a fixed tibia. The ACL, one of the most commonly

The Problem of Falls in Older Adults

Demographic trends point to an exploding older population. Health problems facing the elderly will become even more of a public health issue in the decades to come. One of the most common problems is falls, which often result in injury and hospitalization. Fall-related injuries can be especially devastating because they frequently lead to functional dependence and decreased quality of life.

Many factors contributing to falls have been identified. These include declines in strength, balance, coordination, visual ability, proprioception, and reaction time. Training programs addressing these factors hold promise for decreasing the risk of falls in older people.

injured ligaments in the body, suffers damage when the knee experiences extreme valgus-rotation loads or is violently hyperextended.

An ACL-deficient knee results in altered gait mechanics and various compensatory adaptations. Individuals with ACL deficiency commonly adopt a **quadriceps avoidance** gait pattern. The quadriceps muscle group, acting through the knee extensor mechanism, tends to pull the tibia anteriorly. One of the primary functions of an intact ACL is to restrict anterior translation of the tibia relative to the femur. In the case of a partially or completely torn ACL, the quadriceps avoidance strategy results in decreased activation of the quadriceps and an accompanying decrease in knee joint moment. This reduces the anteriorly directed forces on the tibia and effectively provides compensatory stabilization.

The third condition involves gait pathologies in children with **cerebral palsy** (CP). Cerebral palsy is a nonprogressive condition of muscle dysfunction and paralysis caused by brain injury at or near the time of birth. CP affects more than half a million Americans. There are many forms of cerebral palsy, the most common being spastic diplegia. In **spastic diplegia**, the child's posture and gait are characterized by abnormally flexed, adducted, and internally rotated hips; hyperflexed knees; and **equinus** of the foot and ankle.

In addition to the mechanical deviations characteristic of CP gait, walking exacts a much greater physiological demand on children with CP. One study found 63% greater oxygen consumption (relative $\dot{V}O_2$) in children with spastic CP compared with controls. This increase was largely caused by inefficient energy transfer between and within adjacent body segments during the gait cycle (Unnithan, Dowling, Frost, & Bar-Or, 1999).

Treatment approaches for CP-related gait pathologies include neuromuscular training, orthotics (e.g., ankle–foot orthotics, or AFOs), drugs, and surgery.

concluding comments

This chapter makes clear that the fundamental movements we use on a daily basis are complex phenomena requiring sophisticated integration of our body systems with the external environment. "A conclusion that seems inescapable is that each of us learns to integrate the numerous variables that nature has bestowed on our individual neuromusculoskeletal systems into a smoothly functioning whole" (Inman, Ralston, & Todd, 1994, p. 3).

critical thinking questions

1. Describe why posture and balance are crucial for coordinated human movement.
2. Explain the differences between static posture and dynamic posture.
3. Describe the factors that contribute to balance impairment in older adults.
4. Explain how poor or incorrect posture during sitting, standing, and movement can increase the risk of injury.
5. Define a gait cycle for walking, and outline the subdivisions of stance and swing.

6. What is the difference between a step and a stride? Indicate how these variables can be used to determine walking velocity.

7. Explain the muscle activity at the hip, knee, and ankle throughout the gait cycle, including why and when a specific muscle is active.

8. Outline the initial stages of infant locomotion up to the time of walking. Include an approximate time line with your answer.

9. Numerous factors can produce dysfunctional, or pathological, gait. Describe Trendelenberg gait and quadriceps avoidance gait, including the anatomical dysfunctions producing the walking pattern.

suggested readings

Cech, D.J., & Martin, S.M. (2002). *Functional movement development across the life span*. Philadelphia: Saunders.

Enoka, R.M. (2002). *Neuromechanics of human movement* (3rd ed.). Champaign, IL: Human Kinetics.

Haywood, K.M., & Getchell, N. (2001). *Life span motor development*. Champaign, IL: Human Kinetics.

Inman, V.T., Ralston, H., & Todd, F. (1981). *Human walking*. Baltimore: Williams & Wilkins.

Levangie, P.K., & Norkin, C.C. (2001). *Joint structure and function: A comprehensive analysis* (3rd ed.). Philadelphia: Davis.

Neumann, D.A. (2002). *Kinesiology of the musculoskeletal system: Foundations for physical rehabilitation*. St. Louis: Mosby.

Nordin, M., & Frankel, V.H. (2001). *Basic biomechanics of the musculoskeletal system*. Philadelphia: Lippincott Williams & Wilkins.

Perry, J. (1992). *Gait analysis: Normal and pathological function*. Thorofare, NJ: Slack, Inc.

Rose, J., & Gamble, J.G. (Eds.). (1994). *Human walking*. Philadelphia: Williams & Wilkins.

Smith, L.K., Weiss, E.L., & Lehmkuhl, L.D. (1996). *Brunnstrom's clinical kinesiology*. Philadelphia: Davis.

Whittle, M.W. (1996). *Gait analysis: An introduction*. Oxford, UK: Butterworth-Heinemann Medical.

nine

Fundamentals of Running, Jumping, Throwing, Kicking, and Lifting

objectives

After studying this chapter, you will be able to do the following:

- Describe the movement characteristics and muscle control of fundamental sport skills

- Explain how these movements are affected by age and skill

- Give examples of injuries common to these movement skills

> We run, not because we think it is doing us good, but because we enjoy it and cannot help ourselves....The more restricted our society and work become, the more necessary it will be to find some outlet for this craving for freedom. No one can say, 'You must not run faster than this, or jump higher than that.' The human spirit is indomitable.
>
> —*Sir Roger Bannister*

Each sport has its own unique set of movement skills. However, some fundamental sport-related skills are found across many different sports. These fundamental sport skills include running, jumping, throwing, kicking, and lifting. Each of these basic skills is examined here in terms of its movement characteristics and muscle control.

Running

As we walk faster and faster, there comes a point when walking becomes awkward and uncomfortable, and continued increases in speed require us to shift from walking to a running gait.

Why do we run? Usually, to move faster. Whether we are late for class or an important meeting, or taking advantage of the increased metabolic demands of running to increase our level of physical conditioning, or sprinting to win a race, we adopt a running gait to move faster.

The transition from walking to running typically happens at about 2 m/s, or 4.5 mph, and in most cases running is faster than walking, but speed is *not* the distinguishing characteristic between these two gait modes. Race walkers, for example, can reach walking speeds of more than 12 km/h (7.45 mph) (Murray, Guten, Mollinger, & Gardner, 1983), while a slow jogger runs at a much slower speed. Instead of speed, the defining difference between walking and running is that walking has periods when both feet are in contact with the ground (which running does not), and **running** has a period of nonsupport when both legs are in the air (which walking does not).

The definition of running is simple. The execution of running is not. As with walking, running requires a complex interaction of many body systems. And as even casual observation of runners makes clear, we do not all run in the same way. Superb athletes run with remarkable beauty and effectiveness; a person with injury or disease will run much more slowly; infants and older adults run much differently than do young adults.

Each person's running style depends on many factors, including gender, size, strength, balance, anatomical structure, level of conditioning, and skill. These factors result in a wide range of running styles and considerable variability among individuals. Although many of the aspects of running discussed in this section may seem to imply a uniform running style, always keep in mind that each of us is different, and we run in different ways.

> Now here, you see, it takes all the running you can do, to keep in the same place. If you want to get somewhere else, you must run at least twice as fast as that!
>
> —Lewis Carroll (1832-1898)
> Alice's Adventures in Wonderland

Gait Cycle

The running gait cycle differs from that of walking. As explained earlier, walking is characterized by alternating periods of single support, when one of the legs is in its swing phase while the other singly supports the body's weight, and brief periods of double support, during which both feet are in contact with the ground.

Running, in contrast, does not have a period of double support. Rather it has alternating periods of single support, separated by a **flight phase** when both feet are airborne. This flight phase is also referred to by other authors as the *nonsupport phase* or **float phase**. The phases of the running gait cycle are shown in figure 9.1.

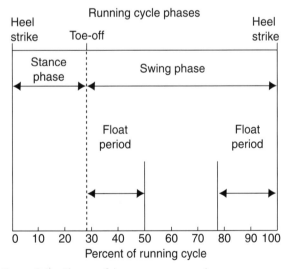

Figure 9.1   Phases of the running gait cycle.

Adapted, by permission, from P.K. Levangie and C.C. Norkin, 2001, *Joint structure and function*, 3rd ed. (Philadelphia, PA: F.A. Davis Company), 470.

Temporal and Spatial Characteristics of Running Gait

Phase timing and stride lengths during running change as speed increases. As gait velocity increases from slow to fast walking, and then to running, the duration of the stance phase decreases, while the duration of the swing phase changes little. With increases in speed, the stance–swing ratio changes from the 60:40 of normal walking to 20:80 in sprinting at 9.0 m/s (20 mph) (figure 9.2).

As with walking, running speed is also calculated as the product of stride length and stride rate (equation 8.1, page 170). Initial increases in speed are made primarily by lengthening the stride. Velocity increases at higher speeds are achieved by quickening the stride rate, largely because there is an anatomical limit to stride length.

Muscle Activity and Control

Given the kinematic differences between walking and running, it is not surprising that electromyographic (EMG) patterns differ as well.

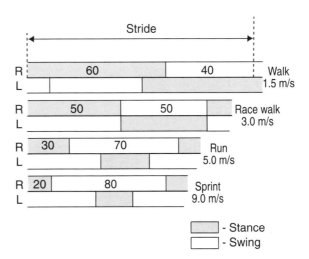

Figure 9.2 Stance-swing ratio changes from walking to sprinting.

Adapted, by permission, from C.L. Vaughan, 1984, "Biomechanics of running gait." *CRC Critical Reviews in Biomedical Engineering* 12:6.

At the hip, the gluteus maximus is active at the end of the swing phase, just before foot contact, to eccentrically slow hip flexion. Similarly, the hamstring group (semitendinosus, semimembranosus, biceps femoris) is active in late swing to slow knee extension before contact.

During early stance, the gluteus maximus, quadriceps group (vastus medialis, vastus lateralis, vastus intermedius, rectus femoris), and gastrocnemius actively generate extensor moments at the hip, knee, and ankle, respectively. These extensor moments provide support and resist the tendency of the ground reaction force to flex the lower-extremity joints.

The plantar flexors, especially the gastrocnemius, are active in late stance to produce the propulsive force necessary for effective push-off into the swing phase. The hamstrings are active to produce the knee flexion during swing. In the swing phase, the rectus femoris is active to aid with both hip flexion and knee extension.

The tibialis anterior shows activity through much of the running gait cycle. During swing, the tibialis anterior acts concentrically to dorsiflex the ankle, while in early stance it is coactive with the plantar flexors to stabilize the foot and ankle.

Increases in running speed are accompanied by greater EMG peak activity and overall levels of activation, shorter *absolute* periods of activity (due to shorter gait cycle duration), but higher *relative* duration as a percent of the gait cycle (Mero & Komi, 1986).

How Fast?

How fast can we run? The winners of the 100 m dash in the Olympic Games are recognized as the world's fastest man and woman. The winning times at the 2004 Olympics in Athens, Greece, were 9.85 seconds by Justin Gatlin (U.S.) in the men's competition and 10.93 seconds by Yuliya Nesterenko (Belarus) on the women's side. These winning times are just ticks slower than the world records as of mid-2005: 9.77 seconds for men (Asafa Powell, Jamaica, 2005) and 10.49 seconds for women (Florence Griffith-Joyner, U.S., 1988). By the time you read this information, these remarkable sprint times may have been eclipsed.

The peak speed in a 100 m sprint is not reached until about 50 to 60 m into the race. From that point on, the sprinter's speed typically decreases a bit. The winner, therefore, is not necessarily the fastest runner at a given moment but rather the one who can maintain speed (i.e., "hold on") through the finish line.

The highest measured human peak running speed is just under 28 mph (just over 12 m/s). Compare this with the fastest land animal, the cheetah, whose top speed can exceed 70 mph. One of the slowest is the snail, who pokes along at 0.03 mph.

Life-Span Perspective

In comparison with walking, running has a shorter life-span perspective, simply because infants lack the strength to propel themselves into the air and thus cannot begin to run until well into their second year or later, and most older adults no longer run, either by choice or because of limitations imposed by injury, disease, or declines in strength and endurance. As a result, most of the research literature on running focuses on children and younger adults.

Infants and Children

Until infants develop sufficient muscle strength to produce a flight phase, they are limited to walking gait. When muscular development allows force production great enough to push the body off the ground, the infant begins development of a running gait. This typically occurs during an infant's second year but can vary from one infant to another.

Early running is characterized by short steps and brief flight periods. As strength and balance improve, running gait matures and becomes more like that of an adult, with increases in stride length, joint range of motion, and trunk rotation, along with a narrower base of support, less lateral limb movement, and coordinated action between the arms and legs. Although these changes begin early in childhood, complete maturation of running gait may not occur until children reach their teenage years.

Cech and Martin (2002) describe the following developmental levels of running:

Level 1

- Upper extremities: Arms are held high to assist with balance control but otherwise are not active.
- Lower extremities: Feet are flat; minimal flight, with swing leg slightly abducted.

Level 2

- Upper extremities: Arms begin to swing as trunk rotation counterbalances pelvic rotation; arms may appear to "flail."
- Lower extremities: Feet remain flat and may support knee flexion more during weight transfer; longer flight phase.

Level 3

- Upper extremities: Arm swing increases in response to trunk rotation.
- Lower extremities: Heel contact made at foot strike; swing leg is in the sagittal plane; support leg reaches full extension at toe-off.

Level 4

- Upper extremities: Arm swing becomes independent of trunk rotation; arms move in opposition to one another and contralateral to leg swing.
- Lower extremities: Similar to level 3.

Each progressive level depends on the developmental status of the child. Increases in size and strength, coupled with neuromuscular system development, allow the child to adopt more complex movement patterns and eventually reach a mature running gait.

Older Adults

Some older adults continue to run well into their later years. Remarkable feats of running prowess are not uncommon. American Bill Galbrecht, for example, completed marathons on each of the seven continents between 1997 and 1999, finishing the last race at age 71 (Cunningham, 2002). By that age, most people have given up running in favor of less-demanding fitness activities such as walking and swimming.

Those who do continue to run in their later years may show some changes in their temporal and spatial characteristics, often due to declines in strength, balance, and joint range of motion. Older adults typically jog and run more slowly, show less knee flexion during swing, and have shorter stride lengths than do younger runners.

Shin Splints: A Case of Vague Identity

"Of the many catchall terms used in the medical literature, perhaps none can match *shin splints* when it comes to nonspecificity, lack of consensus on meaning, and being a continuing source of misunderstanding and confusion" (Whiting & Zernicke, 1998, p. 165). O'Donoghue (1984, p. 591) astutely notes, "As with many names in common use, there is considerable and often heated argument as to what is actually meant by the term. As is usual in these circumstances, the term 'shin splints' is a wastebasket one including many different conditions. The authors of various articles on the subject are inclined to state very definitely that it is caused by one particular thing to the exclusion of all others, which causes great confusion." Although many continue to use the term *shin splints*, clarity would be served by instead using terms that are clinically specific.

Running Injuries

Running imposes high forces on the body that can lead to a variety of injuries. Because forces are felt first by the feet, and then progressively at the ankle, knee, and hip, these injuries most often afflict the lower extremities. The response to injury varies. Most injured runners modify their running (e.g., lower mileage) or stop running altogether until the injury heals sufficiently. Some determined runners, however, continue to run while injured. This can worsen the injury or lead to another injury, as when altered running mechanics in response to foot pain causes a compensatory, or secondary, injury at the knee or hip. Highly competitive athletes, such as professionals whose livelihood depends on their continued running, often try to "play through" an injury and train or compete in pain. They do so at their own risk.

Many different injuries are associated with running. Injuries can be *acute* (e.g., twisting an ankle while running on an uneven surface; straining a hamstring muscle during fast running or sprinting) but more commonly are *chronic,* caused by the repetitive ground reaction forces applied to the foot and transmitted up through the joints of the lower extremities. Among the most common chronic running injuries are knee pain (e.g., chondromalacia patella), calcaneal (Achilles) tendinitis, plantar fasciitis, stress fractures (usually of the tibia or metatarsals), iliotibial band syndrome, and so-called shin splints (see accompanying box).

Altered running gait can be caused by physical pathologies or by other factors, whimsically suggested by the following poem, which has been variously attributed to several authors over the years:

> The centipede was happy quite
>
> Until a toad in fun
>
> Said, "Pray, which leg goes after which?"
>
> That worked her mind to such a pitch
>
> She lay distracted in a ditch
>
> Considering how to run.

Jumping

Citius, Altius, Fortius, the motto of the International Olympic Committee, means "swifter, higher, stronger." To reach the second goal of going higher, one must jump.

Jumping means "to spring free from the ground or other base by the muscular action of feet and legs" (Mish, 1997, p. 634). This definition provides a general description of the jumping action but does not distinguish between different ways of launching and landing. To make these distinctions, *jumping* applies to when individuals propel themselves from the ground with one or both feet and then land on both feet. *Hopping* involves propelling from one foot and landing on the same foot. *Leaping* describes the movement when individuals propel from one foot and land on the other foot (Haywood & Getchell, 2001).

Even though these terms are descriptive and specific, they still do not include all forms of jumping. In athletic competition, for example, high jumpers leave the ground from one foot and land in the pit on their backs. They clearly jump, but their actions do not fit into any of the standard definitions.

Types of Jumping

Jumping comes in many forms. Children at play jump out of sheer joy. Athletes jump to grab a rebound in basketball or catch a pass in American football. Ballet dancers jump when performing a grand jeté. Physical education students do jumping jacks. Boxers jump rope. The list goes on and on.

Jumping is also used to test lower-extremity power output (e.g., vertical jump test) and provides performance challenges to see how high (e.g., high jump) and how far (e.g., long jump, triple jump) one can jump. Each jump type has a specific goal and therefore requires a unique set of movements and pattern of muscle involvement.

With so many different types of jumps, it is infeasible to analyze here the joint motions and muscle control of all of them. Thus, we describe a basic standing vertical jump with a two-foot takeoff and landing. The fundamental patterns described here are modified for other jump types, but many of the basic concepts, such as preparatory leg and arm action (i.e., countermovements), apply to most jump types.

A standing vertical jump can be divided into four phases: preparatory, propulsive, flight, and landing (figure 9.3). The jump begins from a normal standing position. During the preparatory (down) phase, the hip and knee joints flex, the ankles dorsiflex, and the arms swing back into hyperextension. In the propulsive (up) phase, the hips and knees extend, the ankles plantar flex, and the arms swing forward in flexion. The flight phase begins at takeoff when the toes leave the ground. Throughout the flight phase, the body assumes a relatively upright posture that is maintained until landing. Following flight, first ground contact begins the landing phase, during which the hips and knees flex, with ankle dorsiflexion and extension of the arms, as the body absorbs the forces of landing.

The proper timing of joint motions is critical for a successful and proficient jump. During the propulsive phase, for example, there is a rapid proximal-to-distal sequencing of maximum angular velocity at the hip, knee, and ankle joints, with very small delays between adjacent segments (Hudson, 1986). This sequencing is necessary for the effective transfer of energy, from one segment to the next, required for optimal jumping performance. Alterations in this sequencing, such as when a jumper is fatigued, can alter the mechanics of the jump and result in a lower jump height.

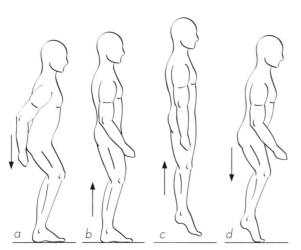

Figure 9.3   Phases of a standing vertical jump: *(a)* preparatory, *(b)* propulsive, *(c)* flight, *(d)* landing.

Muscle Activity and Control

We know from our discussion in chapter 4 that actively stretching a muscle before its active shortening enhances the muscle's force capability. This eccentric–concentric stretch–shorten cycle is used to good advantage in jumping. Nearly a century ago, researchers demonstrated that a jumper performing a countermovement jump (i.e., with a preparatory squat immediately before the upward propulsive phase) could jump higher than when executing a static jump (i.e., a jump begun from a squat position with no preparatory downward movement). One of the primary reasons for the better performance is that muscles during the preparatory phase act eccentrically immediately before their concentric action during the propulsive phase.

More specifically, hip and knee extensors (e.g., gluteus maximus at the hip and the vasti muscles at the knee) and ankle plantar flexors (e.g., gastrocnemius) act eccentrically during the preparatory phase to control flexion of the lower-extremity joints. When the hip, knee, and ankle joints reverse from flexion (during the preparatory phase) to extension (propulsive phase), the muscle actions also reverse from eccentric to concentric. This creates a classic stretch–shorten cycle for all the involved muscles, thereby facilitating force enhancement and thus a higher jump. And just as we saw a proximal-to-distal sequential pattern in the joint angular velocities, we see a similar pattern in the maximal activation of the hip, knee, and ankle extensors as well.

During the early flight phase, the lower-extremity muscles show markedly lower activity. Toward the end of the flight phase and before ground contact, the hip and knee extensors and ankle plantar flexors show anticipatory activity in preparation for landing. This preactivation is necessary to stiffen the muscles and better prepare them to eccentrically accommodate the high ground reaction forces at impact and early in the landing phase. In the case of an immediate second, or repeated, jump, this eccentric loading allows a smooth transition into the next stretch–shorten cycle for the subsequent jump.

The arms play an important role in successful jumping. They provide balance throughout the jump and add to the energy and momentum of propelling the body upward. Biomechanical studies estimate that the arms contribute 10% to the vertical velocity at takeoff (Luhtanen & Komi, 1978). During the preparatory phase, the arms swing backward into hyperextension through concentric action of the glenohumeral (shoulder) extensors (e.g., posterior deltoid). In the propulsive phase, the arms swing forward through concentric action of the shoulder flexors (e.g., anterior deltoid, pectoralis major).

Jumping is an essential element in many sports and dance forms. In some, the jump itself is the primary focus. For example, in athletic field events such as the long jump, triple jump, and high jump, the goal is to jump as far or as high as possible (figure 9.4). Success in these activities requires considerable speed and coordination.

© Michael Zito/SportsChrome

a   b   c

Figure 9.4   Jumps in athletics (track and field): *(a)* long jump, *(b)* triple jump, *(c)* high jump.

Figure 9.5 Examples of jumping in *(a)* ballet, *(b)* basketball, *(c)* volleyball.

In other activities, jumping plays an important role but is not the sole focus of the activity. Success in ballet, basketball, and volleyball, for example, is enhanced by jumping skills (figure 9.5).

Life-Span Perspective

As with running, the life span of jumping is shorter than that of walking. The explosive strength necessary to propel the body through the air does not develop until the end of a child's second year. At the other end of the life span, older adults, with rare exceptions, abandon jumping as a movement form.

Infants and Children

As children learn to jump, they typically move from simple jumping patterns to more complex ones. Initially, they jump from two feet and land on two feet. Next, they may jump down from one foot to two feet, then jump down from two feet to one foot. This is followed by jumping forward from two feet to two feet and then to running and jumping forward from one foot to two feet. Children later develop the ability to jump over objects and perform more complex jumping patterns that may involve body rotations and more use of the upper extremities. The development of jumping skills begins at about age 2 years, with many of the basic jump types in place by the end of a child's fourth year. Many children do not master jumping skills until much later, with some showing poor jumping technique even in their teens.

Developmental progress in jumping can be assessed by the age at which a child performs a particular jump, the height or distance of a jump, and the jumping form or technique. Proficient jumping is characterized by a preparatory crouch that prestretches the muscles of propulsion, allowing them to generate maximal force as the lower-extremity joints fully extend at takeoff, and a backward arm swing followed by a vigorous forward arm swing to initiate takeoff.

When jumping for maximum height, proficient jumpers direct force downward to the ground; extend the body throughout the flight phase; keep their trunks relatively upright throughout the jump; and flex the hips, knees, and ankles on landing to absorb the ground reaction forces. In jumping for maximum horizontal distance, jumpers lean their trunks forward; direct force downward and backward, with the heels leaving the ground before knee extension during push-off; flex the knees during flight; swing the lower legs forward for a two-foot landing; flex at the hips to assume a "jackknife" position at landing; and flex the hips, knees, and ankles on landing to absorb the contact forces (Haywood & Getchell, 2001).

It is well established that the stretch–shorten cycle (SSC) aids performance in adults by enhancing muscles' ability to generate force. But do children make use of the SSC and, if so, to what extent? To explore this question, Harrison and Gaffney (2001) compared vertical jump performance of countermovement and static jumps in children and adults. Using vertical velocity at takeoff as the criterion measure, they found that children do indeed make use of the SSC, but performance was more variable in children than in adults, suggesting that children perform countermovement jumps in a nonoptimal way.

Older Adults

Older adults typically cease jumping, even earlier than they might abandon running. Many factors contribute to this tendency, including declines in muscle strength and power, compromised balance, increased risk of injury, and fear of falling. As a result, there is scant research describing the characteristics of jumping in older adults. We know little about how the jumping patterns of older individuals compare with those of younger adults.

Jumping Injuries

Most jumping-related injuries involve the knee. The knee joint complex forms the critical middle link in the kinetic chain of the lower extremity. In this role, its loading and motion characteristics dictate effective limb function. The most important component of the knee complex is the **knee extensor mechanism** (KEM), consisting of the quadriceps muscle group, the patellofemoral joint, and the tendon group connecting these elements. The patella serves as the central structure in the KEM. In that role, the patella acts as a pivot to enhance the mechanical advantage of the quadriceps during knee flexion and extension.

In jumping, the KEM is essential for effective performance. Injuries affecting the KEM compromise jumping ability. These injuries include patellar maltracking (described in chapter 3), quadriceps tendinitis (at the superior pole of the patella), patellar tendinitis (at the inferior pole), chondromalacia patella (softening degeneration of the articular cartilage on the back side of the patella), and Osgood-Schlatter disease (inflammation of the bone at the tibial tuberosity where the patellar tendon attaches to the tibia).

Throwing

Throwing is as old as humankind. In prehistoric times, hunters threw rocks and spears at animals in hopes of securing food for survival. Through the millennia, throwing has been an essential combat skill, early on using rocks and primitive weapons and more recently employing destructive implements such as hand grenades. Many contemporary sports include throwing as an essential skill. These include softball and baseball, American football, basketball, and several events in athletics

> Don't throw stones at your neighbors if your own windows are glass.
>
> —Benjamin Franklin (1706-1790)

(i.e., track and field) such as the shot put, discus, and javelin. In noncompetitive situations, throwing sometimes provides nothing more than a pleasant diversion, as when a thrower tries to skip rocks across the still surface of a mountain lake.

Despite the wide range of venues and goals, all throws are similar in that they involve using the upper extremity to launch a handheld object (**projectile**) through the air. The study of projectile motion is called **ballistics**, and throwing is one of several ballistic skills in which force is imparted to an object to project it through the air. Other ballistic skills include kicking and striking.

Throws are categorized according to upper-extremity limb segment motion and the method of imparting force to the projectile. Classifications include *overarm throws, underarm throws, push throws*, and *pull throws* (figure 9.6). Overarm throwing is used, for example, by baseball pitchers and javelin throwers. Softball pitchers employ an underarm throwing motion to deliver the ball to the plate. Shot-putters use a push throw to project the shot, while discus and hammer throwers employ a pull throw to project their respective implements.

Throwing Principles

Throwing depends on a number of principles, including the transfer of momentum in a proximal-to-distal manner to an object held in the hand. As a result the object is thrust, or propelled,

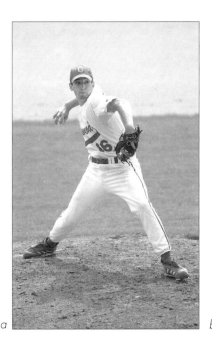

Figure 9.6 Types of throws: *(a)* overarm, *(b)* underarm, *(c)* push, *(d)* pull.

© Empics

into the air. The proper sequencing of limb segment motion presents the neuromuscular system with a challenging muscular control problem. In executing a throw, the body makes good use of the stretch–shorten cycle to enhance force production and throwing distance.

Throwing and Projectile Motion

Projectiles move through the air under the influence of only gravity and air resistance along a path called the **trajectory**. The trajectory is determined by three factors: release height (above the ground), release speed (how fast the object is thrown), and release angle (relative to the horizontal) as shown in figure

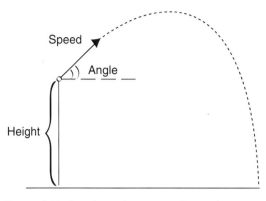

Figure 9.7 Initial conditions at release determine trajectory of a projectile: height, speed, and angle of projection.

9.7. All the thrower's actions before release are intended to produce the proper combination of height, speed, and angle and thereby achieve the throwing goal.

Goals of Throwing

Each throw has a unique goal. Some throwing tasks, such as the shot put and javelin, seek to maximize throwing distance. For distance throws, the most important factor is release speed because speed is more important for distance than are height and angle of release. The goal of other throws may be to maximize height. Still others may have accuracy (e.g., dart throw) or maximization of throwing speed as their goal.

In some cases, a throw may combine goals. An American football quarterback, for example, in attempting to pass the ball to his receiver, must use the right combination of height (to clear the outstretched arms of oncoming defensive linemen), distance (to clear the defensive back), and ball speed (to get the ball to the receiver before another defender can intercept it).

Throwing Phases

The throwing pattern often is divided into phases to facilitate analysis. Each phase has defined beginning and end points, as well as specific biomechanical functions that contribute to the success of the throw.

One general phasing scheme describes three throwing phases: preparation, action, and recovery (Bartlett, 2000). The primary functions of the preparation phase are to (1) put the body in a favorable position for execution of the throw, (2) maximize the range of movement, (3) allow for larger body segments to initiate the throw, (4) actively stretch the agonist muscles to make use of the stretch–shorten cycle, (5) place the muscles at an advantageous length on their respective length–tension curves, and (6) store elastic energy to be used during the action phase.

During the action phase, skillful throwers use sequential muscle actions to execute the throw, beginning with muscles of larger segments, followed rapidly by muscles of the smaller, more distal segments. In most throws, there is proximal-to-distal muscle action and transfer of momentum and kinetic energy. The exact pattern of muscle action and mechanical transfer depends on the goal of the throw.

The primary purpose of the recovery phase is to slow down, or decelerate, the body and its limb segments through eccentric muscle action. This places the body in a favorable balanced position and reduces the chance of injury.

These general phases often are modified, or subdivided, in describing the throwing motion of a particular sport or type of throw. In baseball, for example, the pitching motion typically is divided into five phases: windup, cocking, acceleration, deceleration, and follow-through (figure 9.8). A sixth phase, stride, is sometimes included between windup and cocking. In context of the general scheme just presented, windup and cocking would

Figure 9.8 Phases of baseball pitching: (a-e) windup, (f-h) arm cocking and acceleration, (i-k) arm deceleration and follow through.

constitute preparation, acceleration would correspond with action, and deceleration and follow-through would combine for recovery.

Muscle Activity and Control

Electromyographic studies, using indwelling electrodes, have documented the activity of numerous shoulder, elbow, and forearm muscles during baseball pitching (Gowan, Jobe, Tibone, Perry, & Moynes, 1987; Jobe, Moynes, Tibone, & Perry, 1984; Sisto, Jobe, Moynes, & Antonelli, 1987). During the windup and early cocking phases, minimal shoulder and elbow muscle activity occurs. In the late cocking phase, when the arm approaches maximal external rotation, moderate activity is detected in the biceps brachii. The cocking phase is ended by the pectoralis major and latissimus dorsi, which act eccentrically to slow and eventually stop external rotation at the shoulder. Also active during the cocking phase are the supraspinatus, infraspinatus, teres minor, deltoid, and trapezius. These muscles position the shoulder and elbow for delivery of the pitch. During the acceleration phase, the biceps shows little activity, with propulsion produced by the pectoralis major, serratus anterior, subscapularis, latissimus dorsi, and triceps brachii.

The deceleration and follow-through phases are characterized by eccentric muscle action that decelerates the arm. The biceps, for example, acts eccentrically to slow elbow extension, while glenohumeral external rotators (e.g., infraspinatus, teres minor) actively lengthen to slow the shoulder's internal rotation.

Noteworthy differences showed up in the EMG patterns of professional pitchers compared with less-skilled amateur pitchers. Professional pitchers achieved higher throwing velocities, in part due to stronger shoulder muscle activity (e.g., pectoralis major, serratus anterior, subscapularis, latissimus dorsi). Differences were also evident during the later throwing phases, with the amateurs making greater use of the rotator cuff muscles and biceps brachii (Gowan et al., 1987).

Life-Span Perspective

Many mechanical changes occur in the development of throwing from the first tentative motions of a 2-year-old to the intimidating power of a professional baseball pitcher's fastball. Improved strength, range of motion, and intersegmental coordination all contribute to improved throwing performance.

Infants and Children

When children first throw, they typically use only arm action by reaching back and then extending the elbow. As development progresses, they add trunk action and stepping and use more complex arm action (Roberton & Halverson, 1984).

Trunk action, when present in early throwers, usually involves forward trunk flexion accompanying hip flexion during the throw. Later, the child adds pelvic and trunk rotation, initially as a coupled "block" motion in which the pelvis and trunk rotate together as a single unit. As development continues, the child uncouples the pelvis and upper trunk and begins the forward throwing motion with pelvic rotation (while the trunk remains rotated away from the throwing direction), followed by forward trunk rotation.

Initially, children throw from a standing position with no foot movement. They then throw with an ipsilateral step (i.e., step forward with the foot on the same side as the throwing arm). As development progresses, the child learns to take a short contralateral step and later makes a longer contralateral step.

Throwing development also involves more complex arm action, including a preparatory backswing, alignment of the humerus perpendicular to the trunk, humeral lag (i.e., forward movement of the humerus temporally lags behind trunk and shoulder movement), and forearm lag (i.e., the forearm and hand remain behind until the trunk is facing front).

All these developmental changes allow the young thrower to generate more force and throw the object farther. Developmental pathways share common aspects across all throwers, as just described, but also exhibit individual characteristics unique to each child (Langendorfer & Roberton, 2002).

Older Adults

Mirroring the age-related performance declines in many movement skills, the throwing performance of older adults is similar in terms of velocity to that of children in middle elementary school (Williams, Haywood, & VanSant, 1991). One of the most notable changes is a decline in force production. Even though force production and velocity decline, only small declines in movement form appear evident. Williams, Haywood, and VanSant (1998), in a 7-year longitudinal study of throwing in older adults, report only small declines in movement form and suggest that elderly participants in their study coordinated their movements in a similar way to younger throwers but controlled them differently. Individual cases in the study showed slower movement speeds and decreased range of motion.

Throwing Injuries

In many ways, the human arm is not well designed structurally for repeated, vigorous over-arm throwing. The high number of injuries seen in athletes who throw hard and often (e.g., baseball pitchers, water polo players, javelin throwers) suggests there are limits to the body's ability to withstand the forces of repeated throws. Injuries most commonly happen to the upper extremity but can also afflict the trunk, spine, and lower extremities.

Common upper-extremity throwing-related injuries include impingement syndrome, rotator cuff tears, and medial epicondylitis. **Impingement syndrome** of the shoulder results from repeated arm abduction in which suprahumeral structures (most notably the supraspinatus tendon and the subacromial bursae) are forcibly pressed against the anterior surface of the acromion and the coracoacromial ligament. Overarm throwers are particularly susceptible to impingement syndrome because of the repeated abduction motions inherent to their throwing.

Rupture of musculotendinous structures in the rotator cuff (supraspinatus, subscapularis, infraspinatus, teres minor) is typically the final result of a chain of events that begins with minor inflammation and progresses with continued overuse to advanced inflammation, microtearing of tissue, and finally partial or complete rupture. Compromised tissue integrity and muscle fatigue contribute to altered movement mechanics, and these modified movements further stress the involved tissues and hasten their eventual failure. The supraspinatus is the most commonly injured muscle in the rotator cuff group. Less frequently, other cuff muscles suffer damage. Supraspinatus injury, in particular, is associated with repeated and often violent overhead movement patterns such as throwing.

One of the most common throwing-related injuries at the elbow is **medial epicondylitis**. The wrist flexors in the forearm share a common proximal attachment at the medial epicondyle of the humerus. Eccentric action of these flexors in controlling wrist extension, together with violent valgus-extension loading during the end of the cocking phase, places considerable forces on the medial epicondyle. Repeated loading can lead to inflammation on the medial aspect of the elbow, or medial epicondylitis.

On rare occasions, vigorous throwing can cause bone fractures. Humeral fractures, for example, have been documented as a result of throwing objects as varied as baseballs, javelins, and hand grenades. Various theories have been proposed to explain throwing-related fractures, including factors of antagonistic muscle action, violent uncoordinated muscle action, poor throwing mechanics, excessive torsional forces, and fatigue. Branch, Partin, Chamberland, Emeterio, and Sabetelle (1992) identify additional risk factors in a report on a series of 12 spontaneous humeral fractures in baseball players (average age 36): age, prolonged absence from pitching activity, lack of a regular exercise program, and precursory arm pain.

Although fairly common, throwing injuries are not inevitable. Correct throwing mechanics, proper physical conditioning, and moderation in throwing volume can greatly reduce the risk of throwing-related injuries to the upper extremity.

Kicking

Remember that nobody will ever get ahead of you as long as he is kicking you in the seat of the pants.

Walter Winchell (1897-1972)

Kicking is similar to throwing in that both use proximal-to-distal sequencing to perform the movement. The two skills differ, however, in several ways. Most obviously, kicking is performed by the lower extremity, while throwing involves upper-extremity action. In addition, throwing usually involves an object resting in the hand, which then is launched into the air. In kicking, the lower extremity strikes an object (e.g., a ball) that is not initially in contact with the body. The kicking leg is non–weight bearing throughout the movement as it swings freely; the contralateral leg bears all of the body's weight in supporting the kicking motion.

Success in both skills depends on coordinated action of the entire body and its extremities. Even though the ultimate projection of an object in throwing involves the hand, the lower extremities and trunk also play an important role in a successful throw. In similar fashion, the final contact in kicking is made by the foot, but not without contributions from the upper extremities and trunk.

Kicking is an essential movement in sports (e.g., soccer, American football), various dance forms, and some martial arts (figure 9.9). The plane of action and goal of the kick may differ from one task to another, but there is a common element of a proximal-to-distal segmental progression in swinging the leg forward.

The kicking motion can be divided into phases. The kick usually begins with some type of approach, wherein the kicker moves forward toward the object to be kicked. The purpose of the approach phase is to build momentum that can ultimately be transferred to the object. A preimpact phase follows the approach. This phase begins when the nonkicking leg contacts the ground to provide support for the kick. The support leg blocks the forward motion of the body and helps initiate the thigh swing of the kicking leg. The lower leg (shank) of the kicking leg continues to flex while the thigh begins its forward swing. Immediately before impact, the thigh slows down (decelerates) rapidly, and the shank reverses its direction and extends quickly toward contact.

During the impact phase, the foot is briefly in contact with the object (e.g., ball). The contact time between the foot and the object typically lasts for about 0.1 seconds or less. After impact,

Figure 9.9   Examples of kicking: *(a)* soccer, *(b)* football punt, *(c)* dance, *(d)* martial arts.

the leg continues into a follow-through phase, during which the leg slows down and the kicker regains balance in preparation for the next movement task.

Muscle Activity and Control

The hip joint muscles play a prominent role in any kicking movement. They are responsible for decelerating thigh hyperextension during the approach and reversing its motion to flexion as impact nears. This reversal from eccentric to concentric action allows the hip flexors to make use of the stretch–shorten cycle and produce more forceful action.

Hip muscles are also indirectly involved in knee extension through the whiplike proximal-to-distal sequencing as momentum is transferred from the thigh to the shank and foot. Interestingly, the knee extensors play little part in knee extension during the kick. In a study of soccer kicking, Robertson and Mosher (1985) did not find any knee extensor activity just before ball contact. Instead, they noted knee flexor torques just before contact. These torques may be the body's mechanism to prevent violent hyperextension at the end of knee extension, thereby reducing the chance of a hyperextension knee injury. The authors also suggest that knee extension during the preimpact phase may be too rapid for the knee extensor muscles to keep up with because of limitations dictated by the extensor muscles' force–velocity properties.

Life-Span Perspective

Kicking is a skill used by children and young adults, most often in sports such as soccer and American football. Most of the research on kicking, therefore, focuses on younger populations. Older persons usually do not participate in activities involving kicking. Thus little, if any, research information details the kicking profile of older individuals.

Proficient kicking entails a preparatory windup, sequential segmental movement, a full range of motion at the swinging hip, trunk rotation to improve range of motion, backward body lean at contact, and oppositional use of the arms for balance. Young children initially kick with a simple leg push and exhibit none of the characteristics of skilled kickers. Unskilled kickers typically do not exhibit an approach or preparatory step and often kick with a bent knee. As skill develops, kickers begin to show proficiency, with proper foot placement, increased range of motion, full extension of the kicking leg at contact, trunk rotation, and arm opposition (Haywood & Getchell, 2001).

Children in the initial learning stages find it difficult to control kicking direction. Developmentally, focusing on kicking form, rather than aim, is advisable. Once proper form is achieved, further practice will improve aiming ability. Trying to teach proper aiming strategies too early in the learning process may result in tentative kicking motions without progression toward characteristics of proficient kicking (e.g., full range of motion).

Kicking Injuries

Kicking injuries can result from the mechanics of the kick itself or from contact during the kick (e.g., from an opposing player crashing into the kicker). Given its explosive (ballistic) nature, kicking can cause so-called deceleration injuries, when muscles work eccentrically to slow down a rapidly extending joint such as the knee. If the decelerating muscles (e.g., hamstrings) are overwhelmed by the forces of joint extension, injury may result.

In contact sports such as American football and soccer, injuries may occur when the kicker is hit by another player. In football, the punter is vulnerable to rushing opponents intent on blocking the kick. The supporting (nonkicking) leg is exposed to contact when the kicking leg is extended and the hip flexed.

In soccer, an opponent may hit the kicker and cause injury. One of the most common soccer injuries is knee damage resulting from the violent impact of an opposing defender. Most susceptible to injury are the knee's medial collateral ligament, medial meniscus, and anterior cruciate ligament.

Lifting

Lifting involves grasping an object and moving it to another, often higher, location. Many occupations require lifting as an essential job component. Construction workers, for example, must lift materials such as lumber. Nurses lift patients from their beds. Physical therapists provide lifting support for patients who cannot perform a task by themselves. Athletes lift weights to improve their strength (figure 9.10). And so on.

> There is no exercise better for the heart than reaching down and lifting people up.
>
> —John Andrew Holmes

Lifting Techniques

The technique used to lift an object depends on many factors, including characteristics of the object (e.g., weight, size, shape), lifter (e.g., height, weight, strength, range of motion), environment (e.g., surface conditions), and task (e.g., distance moved, lifting speed). Some lifts are restricted to the sagittal plane, while others involve some rotation, or twisting, of the trunk.

Fundamental lifting postures include the squat and the stoop (figure 9.11). Some lifts are performed using a style in between a stoop and a squat. Traditionally, the squat lift has been recommended over the stoop. This advice, however, is based on the classic "lift with your legs and keep your back straight" adage, which may be overly simplistic. In practice, the stoop lift is used more often than the squat, especially for lifting light objects.

Perhaps the most important factor when performing either lift is to maintain a neutral lumbar spine (i.e., avoid excessive spinal flexion). Keeping the load close to the body reduces the compressive and shear loads on the spine.

Figure 9.10   An athlete lifting a barbell from the ground.

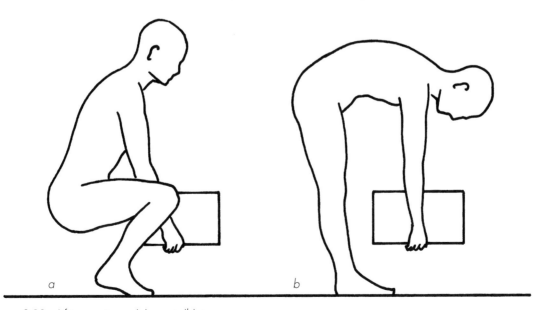

Figure 9.11   Lifting postures: (a) squat, (b) stoop.

Reprinted, by permission, from J. Watkins, 1999, *Structure and function of the musculoskeletal system* (Champaign, IL: Human Kinetics), 155.

Muscle Activity and Control

To lift an object from the ground to a higher level generally involves concentric action of muscles at major joints, along with isometric muscle action for joint stabilization. Controlled lowering of an object (e.g., moving a box from a tabletop to the floor) requires eccentric muscle action. Although lifting heavy objects requires high muscle strength, many lifting tasks involve repeated lifts with lower loads and thus emphasize muscle endurance over muscle strength.

In a traditional squat lift (see figure 9.11a), the lifter primarily uses the major joints of the lower extremity, producing hip and knee extension. At the hip, the gluteus maximus provides the primary hip extensor component. At the knee, the quadriceps muscles (vastus medialis, vastus lateralis, vastus intermedius, rectus femoris) produce knee extension. The degree of arm muscle involvement depends on the task. In sagittal plane lifts above the head, glenohumeral flexors act concentrically to swing the arms forward and upward. The triceps brachii gets involved when elbow extension is required. The stoop lift (see figure 9.11b) involves less leg muscle action and more contribution by the trunk extensors (e.g., erector spinae).

Life-Span Perspective

Most of the research on lifting focuses on healthy adults. Far fewer studies involve lifting in children and older persons. Many of the differences in lifting technique and capability across the life span are dictated by muscular strength and endurance, joint range of motion, and balance.

Infants and Children

Infants and children are limited in what they can lift because of their small stature, short limbs, and limited strength. As they physically mature, their capacity for lifting increases. Because of their greater flexibility and relative body dimensions, children use the squatting technique more often than do adults.

Anticipatory postural adjustments (APAs) are required to perform lifting tasks. Research shows that children are still developing their APA ability at 3 to 4 years, demonstrating inconsistent and immature kinematic and electromyographic patterns in a bimanual load-lifting task (Schmitz, Martin, & Assaiante, 1999). The children in this study also exhibited high intraindividual variability that is expected to decline with age and task mastery.

Older Adults

Lifting ability declines with age, largely because of decreases in muscular strength, range of motion, and balance. Consistent with these declines are changes in lifting strategies based on strength capacity. For example, Puniello, McGibbon, and Krebs (2001) identified three lifting strategies adopted by 91 functionally limited older adults. Subjects with relatively strong hip and knee extensors adopted a strategy dominated by the legs. Those with relatively strong knee extensors but weak hip extensors favored a mixed strategy characterized by initially using the back to lift, followed by the legs. The third approach, a back-dominant strategy, was used by subjects with weak hip and knee extensors. The authors concluded that older adults self-select a lifting strategy based on their hip and knee extensor strength.

Lifting and Low Back Disorders

Low back pain is one of the most costly musculoskeletal disorders in industrial societies. Up to 80% of the population will suffer from low back pain in their lifetimes, and many cases of low back pain result from lifting, especially manual material-handling tasks. The exact cause of back pain often proves elusive. As noted by McGill (2002, p. 118), "Clearly, [low back disorder] causality is often extremely complex with all sorts of factors interacting." Possible mechanisms of low back injury include lifting excessive loads, too many repetitions, and cumulative exposure.

McGill (2002) suggests a two-prong approach for people with low back pain to reduce discomfort and improve function. First, remove the stressors (e.g., heavy loads) that cause or exacerbate damage, and second, engage in activities aimed at building healthy supportive tissues (e.g., muscles, ligaments).

The pain associated with low back dysfunction can arise from many sources, including chemical irritation of tissue due to biochemical events associated with inflammation; stretching of connective tissues such as ligaments, the periosteum, tendons, and the joint capsule; compression of spinal nerves; herniation of intervertebral discs; and local muscle spasms (Whiting & Zernicke, 1998).

Although we will never completely eliminate lifting injuries, we can reduce the risk of lifting-related low back pain through prudent lifting strategies and physical training to improve strength, flexibility, and balance.

Lifting Safety Guidelines

The relationship between lifting mechanics and injury risk is complex because of the many mechanical, physiological, and psychophysical factors involved. The U.S. National Institute for Occupational Safety and Health (NIOSH), recognizing the prevalence and costliness of lifting injuries, issued lifting guidelines in 1981, followed by revised guidelines in 1993. These guidelines provide an equation (Waters, Putz-Anderson, Garg, & Fine, 1993) to calculate lifting limits based on several factors, including object weight, distance of the object from the body, object height, movement distance, angular displacement from the midsagittal plane, lifting frequency, lifting duration, and energy expenditure. Although the NIOSH equation provides one useful component in evaluating lifting dynamics, it applies only to certain two-hand lifting tasks. The equation does not apply to any tasks involving lifting with one hand; lasting more than 8 hours; involving unstable objects; using high-speed motions; or performed while seated or kneeling, while working in a restricted work space, or on slippery floors.

McGill (2002) presents the following detailed guidelines for injury prevention, many of which address lifting-related issues:

- Design work tasks that facilitate variety (i.e., don't do too much of any single thing).
- Avoid a fully flexed or bent spine and rotated trunk when lifting.
- Select a posture to minimize the reaction torque on the low back by keeping the external load near the body.
- Minimize the weight being lifted.
- Do not immediately perform strenuous exertions after periods of prolonged flexion.
- Avoid lifting or spine bending after rising from bed.
- Prestress and stabilize the spine during light lifting tasks.
- Avoid twisting while generating high twisting torques.
- Use momentum when exerting force to lower spinal loads.
- Avoid prolonged sitting.
- Adopt appropriate rest strategies.
- Maintain a reasonable level of physical fitness.

concluding comments

Athletes and performers combine basic skills, such as running, jumping, throwing, kicking, and lifting, into movement forms unique to their sport or activity. A softball player, for example, employs running and throwing as an integral part of her performance, just as a soccer player combines running, kicking, and throwing to play his sport and ballet dancers include elements of running, jumping, kicking, and lifting in their performances. Integration of basic skills forms the foundation for successful performance in all activities.

critical thinking questions

1. Define a gait cycle for running. With respect to the subdivisions of the gait cycle, what are the differences between walking and running?

2. Explain the muscle activity at the hip, knee, and ankle throughout the running gait cycle, including why and when a specific muscle, or muscle group, is active.

3. Define jumping, hopping, and leaping. Give examples of sports that involve jumping, and when appropriate, show how some of the jumps may deviate from the standard definition of jumping.

4. Describe how the stretch–shorten cycle, discussed in chapter 4, can be used to enhance jump performance.

5. Explain the muscle activity at the hip, knee, and ankle throughout the jump cycle, including why and when a specific muscle, or muscle group, is active.

6. List the four principal categories of throws, and provide specific sport-related examples for each one. In addition, describe the three general phases of a throw, including the primary function of each phase.

7. Similar to throwing, kicking can be divided into phases. Give a brief description of, and function for, each phase of a general kick. Use specific sport applications to accompany your explanation.

8. Describe several lifting postural adjustments that should be used to avoid injury. Use specific examples to support your answer.

suggested readings

Cech, D.J., & Martin, S.M. (2002). *Functional movement development across the life span*. Philadelphia: Saunders.

Enoka, R.M. (2002). *Neuromechanics of human movement* (3rd ed.). Champaign, IL: Human Kinetics.

Haywood, K.M., & Getchell, N. (2001). *Life span motor development*. Champaign, IL: Human Kinetics.

Levangie, P.K., & Norkin, C.C. (2001). *Joint structure and function: A comprehensive analysis* (3rd ed.). Philadelphia: Davis.

McGill, S. (2002). *Low back disorders: Evidence-based prevention and rehabilitation*. Champaign, IL: Human Kinetics.

Neumann, D.A. (2002). *Kinesiology of the musculoskeletal system: Foundations for physical rehabilitation*. St. Louis: Mosby.

Smith, L.K., Weiss, E.L., & Lehmkuhl, L.D. (1996). *Brunnstrom's clinical kinesiology*. Philadelphia: Davis.

Williams, K.R. (2000). The dynamics of running. In V.M. Zatsiorsky (Ed.), *Biomechanics in sport: Performance enhancement and injury prevention* (pp. 161-183). Oxford, UK: Blackwell Science.

ten

Analysis of Exercise and Sport Movements

objectives

After studying this chapter, you will be able to do the following:

- Determine the principal muscles used in any weightlifting exercise

- Understand how variations in weightlifting techniques affect muscle recruitment

- Describe the two types of aerodynamic drag that resist a cyclist's forward motion

- Explain the effective and ineffective force components, and describe when the leg muscles are most active, during the pedaling cycle

- Identify the most efficient cadences used by road cyclists and explain why

- Describe how proper seat height affects cycling performance

- Describe the drag forces that resist a swimmer's forward motion

- Define and describe the benefits of sculling

- Describe the propulsive forces produced by the arms during swimming and identify the principal upper body muscles used across all four major swim strokes

- Describe the two principal factors of a stroke that govern swimming velocity

- Describe the kicks used for each stroke

- Define plié and contrast a grand plié and a demi plié

- Describe the muscles active during various dance movements

> Nothing is more revealing than movement.
>
> —Martha Graham (1893-1991)

n the previous two chapters we explored fundamental movements. In this chapter, we look at additional movements and related topics. To cover all exercise and sport movements is beyond the scope of this book. We therefore present sections here on resistance training, cycling, swimming, and dance, as exemplar activities, highlighting muscle involvement and movement mechanics for each activity.

Resistance Training

In designing any type of **resistance training** program, we should select exercises that target specific body regions or muscles. One of the many goals in studying musculoskeletal anatomy, therefore, is to analyze movement and determine what muscles are being used, as explained in chapter 7. The muscle control formula in that chapter facilitates movement analysis by outlining a step-by-step procedure for identifying the movements at relevant joints, the type of muscle action (i.e., concentric, isometric, eccentric), and the muscles being trained. Although the formula enables us to determine the principal muscles being used, it does not tell anything about the level of muscle recruitment because that depends on many factors, including joint range of motion, technique, movement speed, and the external resistance being lifted.

> You know you've got to exercise your brain just like your muscles.
> —Will Rogers (1879-1935)

Table 10.1 lists some of the most effective and popular resistance training exercises used to train different regions of the body. The exercises can be performed with dumbbells, barbells, or machines or simply using one's own body weight. Most of the exercises, however, require external weights. Table 10.1 also includes the principal concentric joint movements needed to perform each exercise.

By using the muscle control formula presented in chapter 7 and the information in table 10.1, we explore the primary muscle groups used when performing the following exercises:

Wide-grip lat pull-down

Seated front barbell press

Seated pulley row

Squat

The following examples list the muscles being trained by groups (e.g., shoulder abductors, elbow flexors, knee extensors). For each of these four exercises, and for all the exercises analyzed in table 10.1, challenge yourself to identify the specific muscles being used. Check your answers by reviewing tables 5.2 through 5.6 (pages 87-93).

Table 10.1   Effective Popular Resistance Exercises Used to Train the Major Regions of the Body

Muscle group	Exercise	Concentric joint movement
Forearms	Wrist curls Reverse wrist curls	Wrist flexion Wrist extension
Upper anterior arms	Standing dumbbell curls Hammer dumbbell curls Barbell curls Reverse curls Standing EZ-bar curls Concentration dumbbell curls Preacher curls	Elbow flexion

Muscle group	Exercise	Concentric joint movement
Upper posterior arms	Triceps extension Triceps kickbacks Triceps press-downs Triceps dips	Elbow extension
Shoulders	Front press Dumbbell press	Shoulder girdle upward rotation, glenohumeral abduction, and elbow extension
	Standing dumbbell side laterals	Shoulder girdle upward rotation (only partial movement) and glenohumeral abduction
	Bent-over lateral raises Pec dec rear deltoid laterals	Glenohumeral horizontal abduction (extension) and shoulder girdle adduction
	Front raises	Glenohumeral flexion
Chest	Flat bench press Push-ups	Glenohumeral horizontal adduction (flexion), elbow extension, and shoulder girdle abduction
	Incline bench press	Glenohumeral horizontal adduction (flexion), elbow extension, shoulder girdle abduction, and glenohumeral flexion
	Parallel bar dips	Glenohumeral flexion, elbow extension, and shoulder girdle depression and abduction
	Flat dumbbell flys Cable crossover flys	Glenohumeral horizontal adduction (flexion), shoulder girdle abduction
Back	Close-grip lat pull-downs Chin-ups	Shoulder girdle depression, glenohumeral extension, and elbow flexion
	Wide-grip lat pull-downs	Shoulder girdle downward rotation, glenohumeral adduction, and elbow flexion
	Seated pulley rows	Vertebral column extension, shoulder girdle adduction, glenohumeral extension, glenohumeral horizontal abduction (extension), and elbow flexion
	Back extensions	Vertebral column extension and hyperextension
Legs	Wide-stance squats	Hip extension, hip adduction, and knee extension
	Narrow-stance squats Leg press Lunges	Hip extension and knee extension
	Leg extensions	Knee extension
	Leg curls	Knee flexion
	Standing calf raises	Ankle plantar flexion
Abdomen	Crunches Reverse crunches Stability ball crunches	Vertebral column flexion
	Bicycle maneuver	Vertebral column flexion and vertebral column rotation

Each exercise includes the principal joint movements needed to perform the concentric phase of the motion.

Wide-Grip Lat Pull-Down

The concentric phase of the wide-grip lat pull-down (figure 10.1) consists of elbow flexion, glenohumeral adduction, and primarily downward rotation of the shoulder girdle. The principal muscles being trained, therefore, are the elbow flexors, glenohumeral adductors, and shoulder girdle downward rotators. Using tables 5.4 and 5.5 (pages 90-92), we see that the following muscles are being trained:

Elbow flexors	Shoulder adductors	Shoulder girdle downward rotators
Biceps brachii	Latissimus dorsi	Rhomboids
Brachialis	Teres major	Levator scapulae
Brachioradialis	Pectoralis major, sternal portion	Pectoralis minor

Figure 10.1   Wide-grip lat pull-down.

Keep in mind that the muscles that *produce* the concentric motions are the same muscles that *control* the eccentric motions. In the return to the starting position, elbow extension is controlled eccentrically by the elbow flexors, shoulder abduction is controlled by eccentric action of the shoulder adductors, and shoulder girdle upward rotation is controlled eccentrically by the shoulder girdle downward rotators.

Seated Front Barbell Press

The concentric phase of the front barbell press (figure 10.2) consists of elbow extension, shoulder abduction, and shoulder girdle upward rotation. The principal muscles being trained, therefore, are the elbow extensors, shoulder joint abductors, and shoulder girdle upward rotators. Using tables 5.4 and 5.5 (pages 90-92), we see that the following muscles are being trained:

Elbow extensors	Shoulder abductors	Shoulder girdle upward rotators
Triceps brachii	Middle deltoid	Trapezius
Anconeus	Supraspinatus	Serratus anterior
	Anterior deltoid	
	Biceps brachii, long head	

Figure 10.2   Seated front barbell press.

As the weight is lowered to the starting position, elbow flexion is controlled eccentrically by the elbow extensors, shoulder adduction is controlled by eccentric action of the shoulder abductors, and shoulder girdle downward rotation is controlled eccentrically by the shoulder girdle upward rotators.

Seated Pulley Row

The concentric phase of the seated pulley row (figure 10.3) consists of elbow flexion, shoulder extension, shoulder girdle adduction, and vertebral column extension. The principal muscles being trained, therefore, are the elbow flexors, shoulder extensors, shoulder girdle adductors, and vertebral column extensors. Using tables 5.4 through 5.6 (pages 90-93), we see that the following muscles are being trained:

Elbow flexors	Shoulder extensors	Shoulder girdle adductors	Vertebral column extensors
Biceps brachii	Latissimus dorsi	Rhomboids	Erector spinae
Brachialis	Teres major	Middle trapezius	
Brachioradialis	Pectoralis major, sternal portion		
	Posterior deltoid		
	Triceps brachii, long head		

Figure 10.3   Seated pulley row.

In the return to the starting position, elbow extension is controlled eccentrically by the elbow flexors, shoulder flexion is controlled eccentrically by the shoulder extensors, shoulder girdle abduction is controlled eccentrically by the shoulder girdle adductors, and vertebral column flexion is controlled eccentrically by the spinal extensors.

Squat

In contrast to the previous three examples, each of which began with its concentric phase and returned using eccentric actions, the squat begins with an eccentric phase during descent, followed by concentric muscle action during the ascent (figure 10.4). Because the concentric phase of the squat consists of hip extension and knee extension, the principal muscles being trained are the hip and knee extensors. Using table 5.2 (pages 87-88), we can see that the following muscles are being trained:

Hip extensors	Knee extensors
Gluteus maximus	Vastus lateralis
Semimembranosus	Vastus intermedius
Semitendinosus	Vastus medialis
Biceps femoris, long head	Rectus femoris
Adductor magnus, posterior fibers	

Figure 10.4   Squat.

In the downward phase of the squat, hip flexion is controlled eccentrically by the hip extensors, with knee flexion controlled eccentrically by the knee extensors.

Variations in Technique

How do variations in technique affect muscle recruitment? For example, what principal differences exist between the seven biceps curl exercises listed for the upper arm in table 10.1?

Because elbow flexion is the concentric motion for each exercise, the elbow flexors are the principal muscles used. The position of the forearm, however, does affect recruitment of the biceps brachii and brachioradialis. For example, hammer curls and EZ-bar curls maximize the contribution from the brachioradialis because the brachioradialis moves the forearm to midposition (between supination and pronation). Supinated curls and EZ-bar curls target the biceps brachii because this muscle is also the strongest supinator of the forearm. In contrast, the brachialis is the prime mover for reverse curls because its insertion on the ulna means its line of pull, and therefore its length, is unaffected by forearm pronation or supination. Although the biceps brachii is recruited in a reverse curl, it is at an anatomical disadvantage to maximize its force contribution. This explains why you can't pronate (reverse) curl as much weight as you can during a hammer, EZ-bar, or supinated curl.

Concentration curls and preacher curls help isolate the elbow flexors because the upper arm is supported, thereby reducing the contribution from the shoulder flexors. Next time you perform a standing curl, be conscious of the contraction produced by the anterior deltoid and the clavicular portion of the pectoralis major. Why are these muscles active? As soon as you start to flex the elbow, the weight wants to drive the shoulder joint into hyperextension to balance the load against your body. In other words, the shoulder flexors contract to counteract the extensor torque produced by the weight.

What is the principal difference between a flat and an incline bench press, and why can you lift more weight when performing a flat bench press? For a flat bench press, the primary concentric shoulder joint motion is horizontal adduction, but with an incline bench press, the concentric motion becomes a combination of horizontal adduction and flexion. Because the sternal portion of the pectoralis major is a shoulder joint extensor (once the arm is flexed in front of the trunk), it will start to drop out (i.e., become less active) as the incline increases. Using the incline bench, therefore, targets the clavicular portion of the pectoralis major because it functions as both a horizontal adductor and flexor of the shoulder. Notice that as the incline steepens to 90°, you switch from performing a bench press to performing a shoulder press. The steeper the incline, therefore, the more you lose the sternal portion of the pectoralis major, and the less weight you can lift.

What is the principal difference between a narrow- and wide-stance squat, and why can you lift more weight when using a wide stance? In a narrow-stance squat, the principal concentric hip joint movement is extension. In the wide-stance squat, the concentric hip motion is a combination of extension and adduction. Using this technique, therefore, calls on the hip adductors to work with the hip extensors to control the weight.

Sometimes variations in exercise technique recruit the same principal muscles but for different reasons. For example, both the close-grip and wide-grip lat pull-down recruit the latissimus dorsi, teres major, and sternal portion of the pectoralis major. In the close-grip lat pull-down, these three muscles are recruited as shoulder joint extensors. In the wide-grip lat pull-down, however, they function as shoulder joint adductors.

Although the examples just discussed are specific to resistance training, the basic steps outlined in chapter 7 can be used to analyze any movement. In addition, the exercises in table 10.1, although by no means comprehensive, represent fundamental movements used across all sports and activities and help illustrate the principles of joint movement and muscle function.

Two Common Sport Movements

Entire books (e.g., Hay, 1993) have been devoted to the functional anatomy and mechanics of sports. Given that amount of information, we clearly cannot discuss all sports here. We therefore present two common movement forms—cycling and swimming—to show how the principles of applied anatomy can be used to analyze a given sport or exercise activity.

In Others' Words: John Jerome

The aspect of sports that first rivets our attention is movement: the joyous flow of the human body lifted into fluid motion. Movement is the athlete's principal medium. Movement requires muscles. Muscle, for some reason, gets a bad rap.

The larger culture has a tradition of sneering at muscle—all brawn and no brain, the meathead image of sports. Muscle is a metaphor for the failure of intelligence, the last resort of fools. It is inescapable that the basic stuff—red meat, human skeletal muscle— is the essential currency of sports, no matter what skills and intelligences are thereby ignored. The elements of athletics are effort and motion. Effort and motion, no matter how gross or fine, are produced by muscle. (Jerome, 1980, p. 43)

—John Jerome, The Sweet Spot in Time

Cycling

Cycling is one of the most popular exercise and sport activities, with millions of cyclists pedaling on a regular basis for health benefits and recreation and many others involved in competitive cycling. The sport's popularity is evident in the remarkable worldwide interest in cycling's premier event, the grueling Tour de France. The 3-week tour, held each July in France, is one of the most demanding of all athletic events. Cycling also provides effective exercise for people with injuries or musculoskeletal conditions (e.g., osteoarthritis) that make walking and running painful. In certain populations (e.g., stroke patients or those with spinal cord injury), cycling can also provide rehabilitative benefits.

Aerodynamic Drag

During training and racing, the greatest force resisting the forward movement of the cyclist is wind resistance. The cyclist must work against two types of aerodynamic drag: form drag and surface drag.

Form drag (also referred to as shape, profile, or pressure drag) is produced by the cyclist's body, and the bike, parting the air. The amount of form drag depends on the size, shape, and speed of the cyclist and the bike, as well as the orientation of the cyclist's body with respect to airflow. Using aerodynamic frame tubing, aerodynamic handlebars, and most important, aerodynamic wheels can significantly reduce form drag produced by the bike. In addition, riders can significantly reduce their form drag by assuming the forward lean position. During time trials (when the cyclist races against the clock), the cyclist assumes a position with his back almost parallel to the ground, arms stretched out in front, with elbows and forearms as close as possible while still maintaining control of the bike. This streamlined position not only minimizes the frontal area of the cyclist's body exposed to the wind but also lengthens the principal hip extensors, thereby increasing their force contribution during the pedaling cycle.

Surface drag (also referred to as frictional resistance, skin resistance, skin-friction drag, or skin drag) is produced as air passes over the surface of the cyclist and the bike. For example, friction created between the air and the cyclist's skin and clothing slows the air and, therefore, resists forward motion. Although surface drag is less of a concern than form drag, particularly at high speeds, the use of skin suits, aerodynamic helmets, and aerodynamic shoe coverings are critical for maximizing time trial performance.

Because both form and surface drag increase as the square of the cyclist's forward velocity, small increases in speed, particularly at the high speeds achieved during racing, necessitate an enormous increase in power output by the cyclist. To understand how a cyclist works to overcome air resistance and achieve the high speeds recorded during racing, we need to analyze the pedaling cycle and how the cyclist applies force to the pedals.

Pedaling Cycle

Many sport skills (e.g., rowing, throwing, running, swimming) are divided into phases for ease of analysis—so is cycling, or pedaling. The **pedaling cycle** is normally divided into two phases: a power phase that drives the bike forward and a recovery phase. The power phase starts when the crank arm (the portion of the crank from the crank axis to the pedal) is at top dead center (TDC), or 0°, and ends when the crank arm reaches bottom dead center (BDC), or 180°. Although most of the useful force is applied to the cranks during the power phase, it is possible to apply force to rotate the cranks during portions of the recovery phase. Even for elite cyclists, however, the quantity of force used to propel the bike forward during the recovery phase is small compared with the force produced during the power phase.

Force Application at the Pedals

Because the force applied to the pedals turns the cranks and propels the bike forward, it is important to understand how pedal forces are measured and what they tell us. To record the force produced by the cyclist's legs during the pedaling cycle, researchers use instrumented force pedals to measure the normal and tangential forces acting on the pedal. The **normal force** component acts perpendicular to the pedal surface, whereas the **tangential force** component acts in the anterior–posterior direction along the surface of the pedal (figure 10.5). By measuring the pedal angle relative to the crank arm, the forces acting on the pedal can then be resolved into an effective, or perpendicular, component and an ineffective, or tangential, component acting down the crank arm (figure 10.5). As the name implies, the effective component produces the torque needed to turn the cranks. In contrast, the ineffective component acts parallel to the crank and, as a result, produces no useful external work in propelling the bike.

A clock diagram is a useful way to visualize the force applied to the pedal throughout the pedaling cycle (figure 10.6). For experienced and elite riders, the maximum normal force, and therefore crank torque, usually occurs at approximately 100° after TDC (at approximately 6 on the clock). The clock diagram shows both the magnitude and direction of the applied force at the pedal. Because the muscle actions during cycling are primarily concentric, each muscle's activation pattern helps explain the duration and timing of the forces acting on the crank throughout the pedaling cycle.

Muscle Activation

Figure 10.7 presents the muscle activation patterns typical of well-trained cyclists, measured using electromyography (EMG). As we can see from the EMG records, the principal activity for the hip, knee, and ankle extensors occurs during the power phase, specifically in the first quadrant (90°)

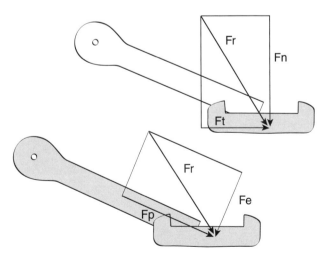

Figure 10.5 Application of pedal forces: resultant force (Fr), normal force (Fn), tangential force (Ft), force parallel to crank arm (Fp), and effective force (Fe).

Adapted, by permission, from P.R. Cavanah and D.J. Sanderson, "The biomechanics of cycling: Studies of the pedaling mechanics of elite pursuit riders." In *Science of cycling* (Champagn, IL: Human Kinetics), 103.

Figure 10.6 Force applied to the pedal during one complete pedaling cycle.

Reprinted, by permission, from P.R. Cavanah and D.J. Sanderson, 1986, The biomechanics of cycling: Studies of the pedaling mechanics of elite pursuit riders. In *Science of cycling*, edited by E.R. Burke (Champaign, IL: Human Kinetics), 105.

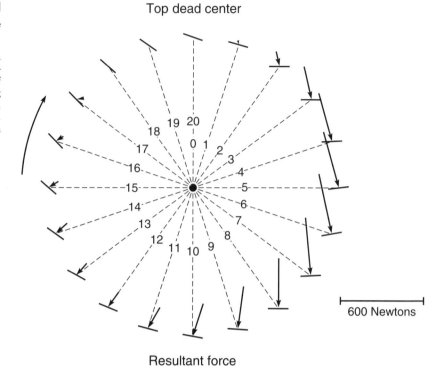

Figure 10.7 Muscle activation patterns for eight leg muscles monitored during steady-state cycling.

Reprinted, by permission, from R.J. Gregor and S.G. Rugg, 1986, Effects of saddle height and pedaling cadence on power output and efficiency. In *Science of cycling*, edited by E.R. Burke (Champaign, IL: Human Kinetics), 74.

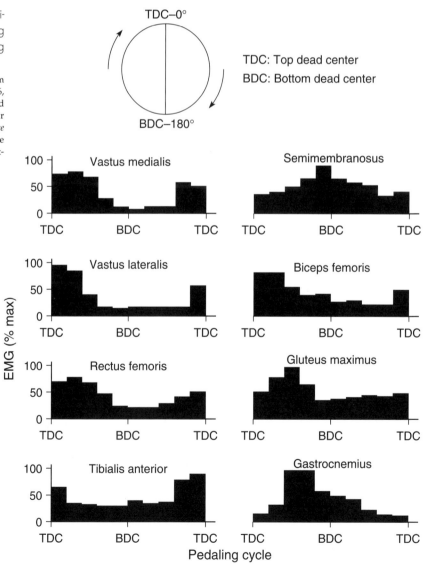

of the pedaling cycle. The high muscle activity recorded from the gluteus maximus, semitendinosus, vastus lateralis, gastrocnemius, and soleus during the initial portion of the power phase explains the timing of peak pedal forces and crank torque around 100° of the pedaling cycle.

The semitendinosus and gastrocnemius activity during the recovery phase primarily results from their role as knee flexors. Their activity helps sweep the leg through BDC and back up toward TDC. In contrast to the three vasti muscles (vastus medialis, vastus lateralis, vastus intermedius), which work only as knee extensors, the rectus femoris tends to be more active during the recovery phase because of its added function as a hip flexor. This hip flexor activity helps bring the recovery leg up to TDC in preparation for the high hip, knee, and ankle extensor activity needed during the power phase of the next cycle. The high muscle activity in the soleus and gastrocnemius during the power phase serves two primary functions: (1) to resist the dorsiflexor torque exerted by the pedal on the ankle and (2) to transmit the large forces, and therefore torques, produced by the hip and knee extensors into the pedals.

Cadence

Most competitive cyclists use a pedaling rate, or cadence, between 70 and 100 revolutions per minute (rpm). Although track cyclists may reach pedaling rates as high as 160 rpm for brief sprints, pedaling rates above 110 rpm are not recommended for road-racing cyclists unless needed for short sprints.

For metabolic efficiency, a slow pedaling cadence should be avoided because of the large energy expenditure needed to maintain prolonged muscle contractions. Although the amount of force increases as the contraction speed decreases, the increased blood vessel occlusion and metabolic energy needed to maintain the contraction could eventually neutralize any advantage of the enhanced force production. If the pedaling rate is too high, little external work can be performed, and energy will be wasted not only in the form of heat but also in overcoming the internal resistance of the muscle.

High cadences and power output elicit an increased reliance on the recruitment of fast-twitch fibers and thus a greater dependence on anaerobic pathways for energy production. The increased blood flow, along with decreased blood vessel occlusion and muscle stress (i.e., less force produced per pedaling cycle) at the higher cadences, however, may help compensate for the greater metabolic cost. In addition, high cadences may produce less metabolic stress on elite cyclists because of their high aerobic capacity. Because the most efficient cadence increases with power output, cadences between 70 and 100 rpm are not excessive for elite cyclists when matched with their high power output during training and racing. The cyclist's optimal cadence, therefore, should produce the most sustainable power for the duration of the event without accumulation of metabolites (e.g., lactic acid) that might impair performance.

Body Position and Seat Height

Although the position of the trunk and arms is critical for aerodynamic efficiency, particularly during cycling sprints, seat height directly affects the cyclist's metabolic and mechanical effectiveness in maintaining high power output. Numerous methods exist for calculating the most effective seat height. Nonetheless, all methods suggest that once the height and fore–aft position of the seat are correctly adjusted, the angle formed between the back of the femur and the leg (shank), with the foot and crank in the BDC position, should be between 150 and 155°. In other words, when the leg is in the BDC position of the pedaling cycle, the knee should be flexed between 25 and 30° (figure 10.8).

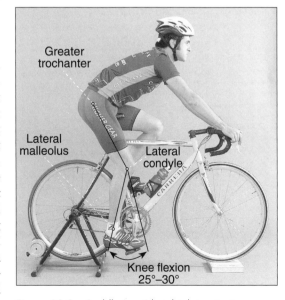

Figure 10.8   Saddle (seat) height determination.

Seat height affects not only the forward lean of the cyclist but also the length–tension and force–velocity properties, and mechanical advantage, of all the principal leg muscles used to propel the bike. No single optimal position, however, exists for all riders. Each rider must rely on continuous testing and trial-and-error experience to ensure maximal performance.

Swimming

Swimming is an excellent exercise for muscular and cardiorespiratory conditioning. It is not effective, however, for increasing bone density because the buoyancy provided by the water negates the impact forces needed for osteogenesis (i.e., bone formation). To increase bone density, weight-bearing activities and resistance training are far more effective.

The water's buoyancy, however, can benefit obese individuals concerned about the risk of musculoskeletal injury associated with land-based activities. Running in water, for example, produces leg and arm motions similar to those of land-based running but without the corresponding joint stress. Studies have also shown that swimming does not produce significant fat or weight loss when compared with land-based weight-bearing exercises. Although the reasons for the negligible fat and weight loss associated with aquatic exercise are unclear, it may be that the body retains the higher levels of fat for temperature regulation and buoyancy.

Resistive (Drag) Forces

Similar to cyclists, swimmers must also contend with drag forces, but in contrast to cyclists, they must overcome the greater viscosity of water. The total drag that resists a swimmer's forward motion is the sum of form drag, wave drag, and surface drag.

Form drag (also referred to as shape, profile, or pressure drag) is produced by the swimmer's body parting the water. The amount of form drag depends on the swimmer's size, shape, and speed and the body's orientation with respect to the flow of water. For both freestyle and backstroke, proper stroking mechanics require the body to rotate around its long axis. Rotating the body not only increases stroke effectiveness but also reduces form drag by decreasing the surface area of the body slicing through the water. To reduce form drag in the butterfly and breaststroke, the swimmer needs to minimize the vertical projection of her head and shoulders to prevent the hips and legs from sinking deeper in the water.

Wave drag is caused by the swimmer's body moving near or along the surface of the water. A portion of the water displaced by the swimmer moves up from a zone of high pressure to a zone of low pressure, creating a wave. Because the swimmer's kinetic energy provides the energy needed to form the waves, the waves present an opposing force to the swimmer. As the speed of the swimmer increases, the opposing force of the waves also increases.

When entering the water after a start or pushing off the wall from a turn, swimmers remain completely submerged in a streamlined position to reduce wave drag. The streamlined position referred to here is the same position used by divers as they enter the water, namely, straight legs with ankles plantar flexed and arms held stretched over the head with hands together. The streamlined position actually serves a dual purpose by decreasing form drag and allowing swimmers to use their forward momentum to propel them through the water. In breaststroke and butterfly, the brief submerging of the body under the water's surface also reduces wave drag.

Surface drag (also referred to as frictional resistance, skin resistance, skin-friction drag, or skin drag) is produced as the water passes over the surface of the swimmer. Friction created between the water and the swimmer's skin and swimsuit slows the water and therefore resists forward motion. Shaving body hair and using new high-tech swimsuits (e.g., "sharkskin" suits) enhance performance by reducing surface drag. Although surface drag is the smallest resistive force encountered by swimmers at high velocities, even a small improvement in performance can make a big difference in top competitions where the margin of victory often is measured in hundredths of a second.

Stroke Mechanics

Similar to cycling, each swimming stroke can be divided into a power and recovery phase. The power phase occurs when the arm pulls through the water and propels the body forward. The recovery phase completes the stroke and repositions the arm for the next power phase.

To maximize forward propulsion during the power phase of all the strokes, swimmers are instructed not to use a straight pull (referred to as **paddling**) with the hand oriented 90° to the surface of the water. Instead, the angle of the hand (angle of attack, or pitch) should vary throughout the power phase, and the hands should follow a curvilinear path through the water (figure 10.9a). This stroke technique is known as **sculling**. In addition, during freestyle and backstroke, the whole body (i.e., shoulders, trunk, hips, and legs) rotates as one unit through a range of 70 to 90°, or 35 to 45° to each side. Controlled by the alternating power and recovery phases of each arm, this rotation reduces form drag, facilitates proper sculling motion of the hand and arm through the power phase, and reduces the muscle effort associated with pulling the arm through the recovery phase.

Propulsive Forces Produced by the Arms and Hands

Although the whole arm pulls through the water during each stroke, it is the hand that produces the largest propulsive force in swimming. The force acting on the hand as it passes through the water is normally divided into two components: a **lift force** that acts perpendicular to the hand's line of pull and a **drag force** that acts in the opposite (i.e., parallel) direction to the motion of the hand. The magnitude of the lift and drag forces is determined by the pitch (angle of attack) of the hand (figure 10.9). How can these forces be used for propulsion? Remember, just because the lift force acts perpendicular to the movement of the hand through the water does not mean that the lift force always acts perpendicular to the surface of the water.

Research shows that the forward propulsive forces produced by lift and drag are greater when the swimmer uses transverse and vertical sculling motions of the hand throughout the power phase of the stroke (figure 10.10 a-c). Because the limb moves backward through the water, the drag force is directed forward, and because the hand is not held perpendicular to the water's surface, a portion of lift can be used for propulsion. Propulsion, therefore, is a continual interaction between the lift and drag forces as the hand follows a curvilinear path through the water. The shape and size of the hand, as well as the speed of the hand moving through the water, affect the magnitude of the propulsive lift and drag forces.

Muscular Control of Stroke Mechanics

The power phase for each stroke can be subdivided into a pull and a push phase. In freestyle, backstroke, and butterfly, the swimmer first pulls his or her body forward over the hands and then continues to push the body past the hands (Colwin, 2002). The pull phase of the

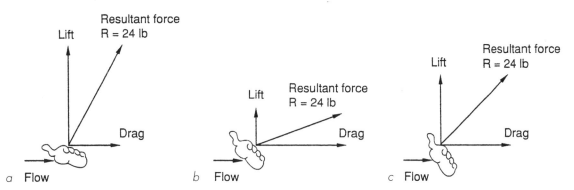

Figure 10.9   Hand angles and resultant force production in swimming: (a) pitch of 15°, (b) pitch of 30°, (c) pitch of 45°.

Adapted, by permission, from C. Colwin, 2002, *Breakthrough swimming* (Champaign IL: Human Kinetics), 38; adapted, by permission, from R.E. Scheihauf, 1977, Swimming propulsion: A hydrodynamic analysis. *In American Swimming Coaches Association world clinic yearbook* (Fort Lauderdale, FL: American Swimming Coaches Association), 53.

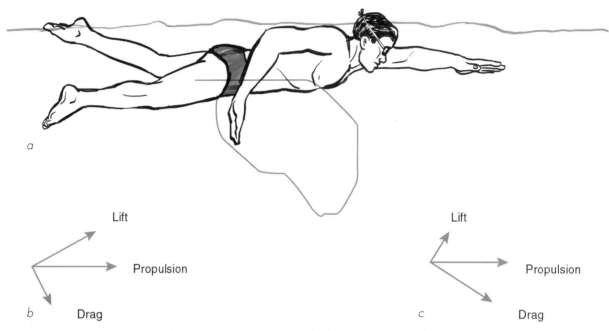

Figure 10.10 *(a)* Swimmer's fingertip trajectory pattern in the freestyle stroke, *(b)* lift-dominated propulsion, *(c)* drag-dominated propulsion.

Reprinted, by permission, from E. Maglischo, 2003. *Swimming fastest* (Champaign, IL: Human Kinetics), 22. Adapted, by permission, from American Swimming Coaches Association, 1977, *American Swimming Coaches Association World Clinic Yearbook 1977* (Ft. Lauderdale, FL: American Swimming Coaches Association), 53.

breaststroke is similar to that of the other strokes, but differs during the push phase in that the body does not pass over the hands.

Even with the added challenge of immersing electrodes in water, muscle activity patterns during swimming have been recorded since the early 1960s. Although the four strokes are significantly different visually, each one uses glenohumeral adduction as a principal joint motion for moving the body forward through the water. The power phase for freestyle, butterfly, and breaststroke also consists of glenohumeral extension, whereas the power phase for backstroke includes glenohumeral flexion. These shoulder joint motions help explain why the primary muscles used across all four strokes are the pectoralis major, latissimus dorsi, and teres major. To assist with the power phase and perform a smooth and powerful recovery, the trapezius, deltoids, elbow extensors, and flexors also play an important role.

Although the longitudinal body rotation characteristic of both freestyle and backstroke is primarily controlled by the alternating stroke mechanics of each arm, the observed rectus abdominis and external oblique activity would appear to help stabilize the trunk and facilitate trunk rotation. Lower limb muscle activity, including the gluteus maximus, semitendinosus, biceps femoris long head, rectus femoris, gastrocnemius, and tibialis anterior have all been recorded during swimming. As expected, when considering the variation across and within kicks, the specific muscle activation patterns can differ substantially.

Swimming Velocity and Momentum

Swimming velocity is largely determined by the product of stroke length (SL) and stroke rate (SR). Depending on the duration and speed of the swimming event, a correct combination of SL and SR must be used to maximize performance. The current trend for skilled swimmers is to increase their velocity by first increasing their stroke length, followed by an increase in stroke rate. Longer strokes help swimmers use the forward momentum generated from their power stroke, thus enabling them to glide more during the stroke. If the stroke rate is too slow, however, the forward momentum will start to dissipate and the swimmer's forward progression will be marked by accelerations and decelerations. In addition to optimizing stroke length and stroke rate, swimmers must maintain a streamlined body position to obtain the maximum benefit from their forward momentum.

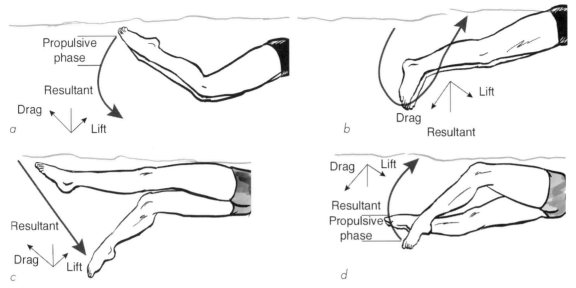

Figure 10.11   Dolphin kick *(a)* downbeat and *(b)* upbeat; flutter kick *(c)* downbeat and *(d)* upbeat.
Reprinted, by permission, from E. Maglischo, 2003, *Swimming fastest* (Champaign, IL: Human Kinetics), 39

Propulsive Forces Produced by the Legs and Feet

Although the thigh, leg, and foot all contribute to the kick for all four strokes, it is the foot that produces the largest propulsive force. When pushing off the wall after a turn or entering the water after a start, the dolphin kick (also butterfly kick) is popular because when combined with a streamlined body position, it allows the swimmer to move with great speed underwater without using the arms (figure 10.11 a and b). Both freestyle and backstroke use the flutter kick for propulsion (figure 10.11 c and d). The flutter kick consists of asynchronous movement of the legs, which contrasts with the synchronous movement of both legs during the dolphin kick. Unlike the other three strokes, the breaststroke uses a propeller-like kick to drive the body forward. Regardless of the kicking technique used, each kick is effective because it produces a net forward thrust force.

Dance

Dancing combines athleticism, grace, and artistic composition. Although dancers can choose to specialize in ballet, modern, aerobic, or jazz dancing, they also have the freedom to combine different dance forms to express their own artistic imagination or the imagination of a choreographer. This great diversity of forms, however, helps explain why dancing has not been the subject of as much research as other movement forms (e.g., walking, running, cycling, swimming). Most of the dance literature and research focuses on technique, biomechanical flaws in technique that may lead to injury, injury rehabilitation, and film or video analysis of movement. Fortunately, more studies are now using electromyographic (EMG) analysis to record and study the way dancers use their muscles to perform basic and complex movements. By documenting the major muscle groups performing fundamental movements, dancers and instructors will be able to diagnose muscle weaknesses, design more effective training programs, and focus on fundamental dance movements to enhance performance and reduce injury.

One of the fundamental dance movements, and perhaps one of the most studied, is the **plié**. The French word *plié* literally means bent or bending. There are two principal pliés: the **grand plié**, or full bending of the knees (the knees should be bent until the thighs are nearly horizontal) (figure 10.12a), and the **demi plié**, or half bending of the knees (figure 10.12b). In addition, both forms of plié should be performed isokinetically during both the lowering, or eccentric, phase and during the rising, or concentric, phase.

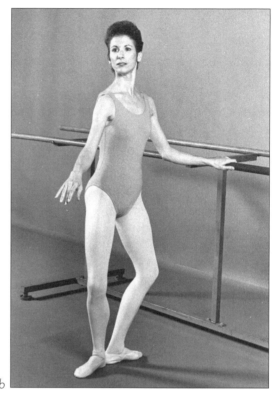

a *b*

Figure 10.12 *(a)* Grand plié and *(b)* demi plié in first position.

In classical ballet, there are five basic foot positions, and every step or movement starts and ends in one of these positions. For example, in first position the hips are laterally rotated, with the heels touching so the feet are parallel with the frontal (coronal) plane. If the grand plié is started with the feet in first position, as in figure 10.12, then proper form requires that the heels lift from the floor during the lowering phase, but not until the flexing of the knees forces the dancer to allow her heels to leave the floor. Likewise, the heels must be lowered to the floor during the rising phase. Every grande plié thus passes through the demi plié position, with the knees half extended and the heels on the floor. In dance training, numerous repetitions of the plié improve strength, flexibility, balance, timing, trunk alignment and stability, and coordination of the joint movements. In addition, the plié is often the first and last element of other movements such as **relevé** (rising on the toes), **pirouettes** (a complete rotation of the body on one foot), and jumps.

Eccentric and Concentric Phases of the Grande Plié

The grand plié consists of two phases: the lowering, or eccentric, phase and the rising, or concentric, phase. The lowering phase can be further divided into two parts: (1) start to heel-off, during which time both the forefoot and heel remain in contact with the floor; and (2) heel-off to midcycle, during which time the heel rises with progressive metatarsophalangeal (MP) joint dorsiflexion (hyperextension). The rising phase is also divided into two parts: (1) midcycle to heel-on, during which time the heel is lowered to the floor; and (2) heel-on to end (i.e., the starting position), during which time both the forefoot and heel are in contact with the floor. Midcycle is the lowest body position during the grand plié, occurring when the dancer reaches maximum hip flexion, hip abduction, knee flexion, and MP dorsiflexion (hyperextension). Maximum ankle dorsiflexion occurs at heel-off during the lowering phase and at heel-on during the rising phase.

Muscle Activation of the Grande Plié

The grand plié closely resembles a resistance training movement similar to a modified squat. Using the same approach we used for resistance training, let's determine the principal muscle groups used during the grand plié. The concentric phase consists of hip extension coupled with hip adduction, knee extension, and ankle plantar flexion. The principal muscles being trained, therefore, are the hip extensors and hip adductors, knee extensors, and ankle plantar flexors. Using tables 5.2 and 5.3 (pages 87-90), we see that the following muscles are being trained during a grand plié:

Hip extensors	Hip adductors	Knee extensors	Ankle plantar flexors
Gluteus maximus	Adductor magnus	Vastus lateralis	Gastrocnemius
Semimembranosus	Adductor longus	Vastus intermedius	Soleus
Semitendinosus	Adductor brevis	Vastus medialis	Peroneus longus
Biceps femoris, long head	Pectineus	Rectus femoris	Peroneus brevis
Adductor magnus, posterior fibers	Gracilis		Tibialis posterior
	Gluteus maximus, inferior fibers		Flexor hallucis longus
			Flexor digitorum longus
			Plantaris

Research by Trepman, Gellman, Micheli, and De Luca (1998) reports activity from muscles associated with each of these groups, including the gluteus maximus, hamstrings, adductors, vastus lateralis and medialis, and gastrocnemius. Muscle activity was recorded from each muscle or muscle group during both the lowering (eccentric) phase and the rising (concentric) phase. Their findings include the following:

- Knee flexion and thigh abduction during the descent are controlled by eccentric action of the quadriceps and adductors, respectively.
- The hamstrings stabilize the hips and knees at midcycle.
- Ascent is produced by action of the quadriceps and adductors, with high levels of muscle activity early in the rising phase.
- Isometric action of the tibialis anterior stabilizes the dorsiflexed ankle during portions of the lowering and rising phases.

Grand plié EMG patterns may be affected by numerous factors including skill level, muscle strength, flexibility, technique (e.g., arm positions, upper body posture, degree of turnout at the hip, and range of motion at each joint), speed of motion, and footwear. Variations in EMG patterns are therefore expected, as they would be in any complex human movement.

concluding comments

The dynamics of the four common forms of exercise (resistance training, cycling, swimming, dance) discussed in this chapter serve as examples of how any physical conditioning, exercise, or sport task can be analyzed. In sports such as cycling and swimming, it is important to study not only the athletes but also the equipment and clothing that directly affect their performance. Although many dance movements require great athleticism, the complex artistic component often makes research difficult.

Having completed this and earlier chapters, you now have the tools to perform fundamental movement analyses on your own. Comprehensive movement analysis encompasses many disciplines, including functional anatomy, biomechanics, physiology, and psychology. By understanding movement from all these perspectives, you have the foundation necessary to pursue advanced work in any of the numerous areas that involve the science of human movement.

critical thinking questions

1. What is the principal benefit of performing concentration and preacher curls instead of the other curling exercises listed in table 10.1? Hint: Why are they considered isolation exercises?

2. Why are you able to lift more weight when performing wide-stance versus narrow-stance squats?

3. Why can you lift more weight when performing a flat versus an incline bench press? In addition, why do you continue to lose strength as the incline of the bench increases?

4. When do competitive cyclists normally produce the maximum effective force during the pedaling cycle? Use your knowledge of the muscle activity patterns during the pedaling cycle to explain your answer.

5. What are the principal benefits of the forward lean position used by cyclists during racing, particularly during time trial racing?

6. Describe the two types of drag force common to both cyclists and swimmers.

7. What are the similarities between a pedaling cycle and a swimming stroke?

8. Explain the importance of the whole-body rotation common to both freestyle and backstroke.

9. What are the principal muscles used during the power phase of each stroke? List the movement functions of each muscle, and explain how these movements relate to each stroke.

10. Swimming velocity is largely governed by stroke length (SL) and stroke rate (SR). Briefly explain the benefit of the current trend for competitive swimmers to first increase their SL and then increase their SR to maximize speed.

11. Explain the muscle activation patterns seen in the grand plié.

suggested readings

Baechle, T.R., & Earle, R.W. (Eds.) (2000). *Essentials of strength training and conditioning* (2nd ed.). Champaign, IL: Human Kinetics.

Bompa, T.O. (2005). *Periodization training for sports* (2nd ed.). Champaign, IL: Human Kinetics.

Bompa, T.O., Di Pasquale, M., & Cornacchia, L.J. (2003). *Serious strength training* (2nd ed.). Champaign, IL: Human Kinetics.

Burke, E.R. (Ed.). (1986). *Science of cycling*. Champaign, IL: Human Kinetics.

Burke, E.R. (2002). *Serious cycling* (2nd ed.). Champaign, IL: Human Kinetics.

Burke, E.R. (Ed.). (2003). *High-tech cycling* (2nd ed.). Champaign, IL: Human Kinetics.

Colwin, C.M. (2002). *Breakthrough swimming*. Champaign, IL: Human Kinetics.

Delavier, F. (2001). *Strength training anatomy*. Champaign, IL: Human Kinetics.

Fitt, S.S. (1996). *Dance kinesiology* (2nd ed.) Schirmer Books.

Fleck, S.J., & Kraemer, W.J. (2004). *Designing resistance training programs* (3rd ed.). Champaign, IL: Human Kinetics.

Hay, J.G. (1993). *The biomechanics of sports techniques* (4th ed.). Englewood Cliffs, NJ: Prentice Hall.

Hollander, A.P., Huijing, P.A., & de Groot, G. (Eds.). (1983). *Biomechanics and medicine in swimming—International series on sport sciences (vol. 14)*. Champaign, IL: Human Kinetics.

Kreighbaum, E., & Barthels, K.M. (1996). *Biomechanics: A qualitative approach for studying human movement* (4th ed.). Needham Heights, MA: Allyn and Bacon.

Laws, K. (2002). *Physics and the art of dance*. New York: Oxford University Press.

Ungerechts, B.E., Wilke, K., & Reischle, K. (Eds.). (1988). *Swimming science V—International series on sport sciences (vol. 18)*. Champaign, IL: Human Kinetics.

Vorontsov, A.R., & Rumyantsev, V.A. (2000). Propulsive forces in swimming. In V. Zatsiorsky (Ed.), *Biomechanics in sport* (pp. 205-231). Malden, MA: Blackwell Science.

Vorontsov, A.R., & Rumyantsev, V.A. (2000). Resistive forces in swimming. In V. Zatsiorsky (Ed.), *Biomechanics in sport* (pp. 184-204). Malden, MA: Blackwell Science.

eleven

Future Directions of Human Movement Studies

objectives

After studying this chapter, you will be able to do the following:

- Identify advances in medicine and technology that affect the study of human movement

- Identify demographic trends and explain how they will create movement challenges in the future

- Describe movement-related social trends that will affect society

- Explain the concept of limits to human performance

> The past is dead, and has no resurrection; but the future is endowed with such a life, that it lives to us even in anticipation.
>
> —Herman Melville (1819-1891)

The study of human movement boasts a rich history that includes fundamental observations over the past two millennia, more rigorous assessment in the last several centuries, and technologically inspired advances in recent decades. In spite of all we have learned, much remains to be discovered. In concluding our exploration of dynamic anatomy and human movement, we move from "what we know" to "what the future might hold." Predicting the future is always tenuous and carries risk of error, of course, but the possibilities of improving our understanding of movement—and applying this knowledge with new technologies to enhance performance, reduce injury, treat movement dysfunction, and improve our quality of life—are most intriguing. Movement-related advances will span many areas, including medicine and technology, and will be influenced by demographic changes and societal trends.

Advances in Medicine and Technology

Physicians face significant challenges in keeping up with technological advances in medical diagnosis and treatment. Developments in medicine and technology are so rapid, and so revolutionary, that medical professionals at all levels find it difficult to keep up to date. On the medical front, recent advances in diagnostic and treatment techniques have been nothing short of miraculous. Many of the procedures we now take for granted, such as routine joint replacement (**arthroplasty**), were inconceivable only a few decades ago. Diagnosis of injury and disease has progressed remarkably through advances in imaging technology, such as computed tomography (CT) scanning, magnetic resonance imaging (MRI), and positron emission tomography (PET) scanning. These imaging technologies allow for the diagnosis of conditions in their early stages when they are most treatable and allow surgeons to view the inside of the body noninvasively.

Noninvasive surgical techniques are now available and promise to grow considerably in the years to come. For example, gamma knife surgery uses finely focused beams of radiation to attack brain tumors without damaging adjacent tissues. This noninvasive technique specifically targets cancerous tissue without traditional surgery and its attendant risks and complications.

Many tenets of medicine once thought to be true are now falling by the wayside as we discover more about the body's amazing ability to repair itself. Patients paralyzed by spinal cord injury, for example, who were thought to be untreatable are now regaining some degree of function, based largely on our improved understanding of the neuromuscular system and advances in treatment modalities.

We are now discovering that body tissues once believed to be irreparable are able to repair or regenerate themselves, at least to some degree. In addition, the field of tissue engineering holds promise of being able to "grow" new tissues and organs using sophisticated techniques involving undifferentiated (stem) cells. The possibility of eventually creating spare body parts to replace defective ones is very real.

With the human genome now well defined, the possibilities of improving overall human function in general, and human movement capabilities in particular, through gene therapy and genetic engineering are virtually endless. The optimism of potential must be tempered, however, by acknowledgment of the risks and uncertainties inherent to any new and developing technology. Potential advances must be made cautiously. Haste, in this case, may make much more than just waste. Bioengineering, for example, holds promise for dramatically improving our quality of life but at the same time carries risks of genetic accidents (e.g., gene manipulation gone awry) and the threat of being used in biological warfare. In any case, medical and technological advances will be influenced by a wide variety of social and economic factors and will inevitably raise moral and ethical concerns.

On the technology front, improvements in computer power and speed, coupled with remarkable miniaturization of computer components (e.g., microchips), open the door to exciting possibilities. Computer chips are already being used to control muscles in patients with spinal cord injury. Chip technology may someday allow implanted computer chips to control many different body organs and eventually treat neuromuscular conditions (e.g., Parkinson's disease) and other debilitating diseases (e.g., Alzheimer's and diabetes).

Technological advances in computer control are also evident in the area of robotics. Computer-controlled robots are now being used both diagnostically, to identify medical conditions and their causes, and surgically. Though still in its infancy, robot-assisted surgery enables surgeons to work with greater precision and in areas difficult to access using traditional methods. Robotics even allows surgeons to service patients in remote locations by controlling surgical tools from a remote site and performing surgery in absentia.

Robotics can also be used to teach surgical techniques in a virtual environment. A medical student can perform surgery on the computer, using robotic and computer technology, with no risk to a patient. Sophisticated simulation instrumentation allows the student to use surgical tools and get the feel of a real surgical procedure while viewing events on a computer screen. Initial learning and later refinement in surgical skills using virtual technology is both cost-effective and safer than traditional methods.

Computer capacity has developed to the point where widespread use of virtual reality is now possible. Virtual reality "is a way for humans to visualize, manipulate, and interact with computers and extremely complex data. . . . Virtual reality represents a major leap in creating ways to interact with computers and to visualize information. Instead of using screens and keyboards, people can put displays over the eyes, gloves on their hands, and headphones on their ears. A computer controls what they sense; and they, in turn, can control the computer" (Aukstakalnis & Blatner, 1992, p. 7). The applications of virtual reality are nearly endless. Among its many applications, virtual reality is used in modeling (by creating three-dimensional models of a room or structure, for example, to allow a person to move virtually around in the structural space), communication (by allowing sharing of information in many locations, in many languages, at the same time), and control (by use of complex computer models to control information and processes). Virtual technologies are currently employed widely in the entertainment field, with uses ranging from video games to computer animation. The tools of virtual reality and computer animation, though still in their infancy, also hold great promise for use in educational and rehabilitation environments.

In Others' Words: Aukstakalnis and Blatner on Virtual Reality

It's been said that members of the medical community are similar to auto mechanics. Certainly both professions specialize in preventative maintenance and in repairing the nonfunctional. And in the pursuit of these objectives, professionals from both camps use a number of tools for analysis and adjustment, not to mention cutting and sewing back together.

But where the drive shaft and transmission mechanisms in a car are always in the same place for a given model, the human body is significantly more variable and complex. Organs like the kidneys and the liver can shift into non-standard places, making them difficult to locate; different people come in different sizes and proportions, sometimes requiring special tools for those who fall outside the range of the average person; and the body is so interconnected that damage in one place may affect the functionality of the entire system.

Thus, where a wrench and a solid understanding of auto mechanics can go far in fixing an engine, the human body generally demands a more complex approach. Virtual reality technology is beginning to be used in various medical and health-related areas and is already showing great promise in becoming a standard tool in some medical processes. (Aukstakalnis & Blatner, 1992, pp. 209-210)

—*Steve Aukstakalnis and David Blatner*, Silicon Mirage: The Art and Science of Virtual Reality

Demographic Trends

Perhaps the most dominant current demographic change is our aging population. With people living longer than ever, the population of those above 65 years of age is predicted to explode in the coming decades. An older population brings with it unique challenges, both to the individual and society as a whole. Many individuals will need to confront the challenges of living well into their 80s, 90s, and beyond. More and more attention and resources must be devoted to ensuring quality of life for older adults. Conditions common in older people, such as Alzheimer's disease, will become more prevalent and place greater demands on our medical and social institutions.

The role of movement in an aging population will take on greater importance. At one level, exercise and physical conditioning programs have already proved effective in enhancing the quality of life in older people. As discussed earlier, some slowing accompanies aging. But as we asked at the outset in chapter 1, do we slow down because we age, or do we age because we slow down? After exploring the amazing ability of the human body to adapt, we must conclude, at least to some degree, that the latter is true. The adoption of sedentary lifestyles as we age contributes greatly to the declining health status of older people.

The encouraging news is that many age-related health conditions are not inevitable. Consider the now-classic study by Fiatarone et al. (1990), who noted significant strength gains after just 8 weeks of high-intensity resistance training in frail, institutionalized persons averaging 90 years. Fiatarone and colleagues demonstrated that high-resistance weight training promoted gains in muscle strength, size, and functional mobility in persons up to 96 years of age! With proper lifestyle choices in the areas of nutrition and exercise—and a little luck—most of us have the capacity to maintain healthy and productive lives well into our later years.

In a broader sense, the role of movement in alleviating age-related declines in health and performance has tremendous implications for society as a whole, both in improving the quality of life for millions of older adults and in reducing the financial and emotional burdens that inevitably accompany disease and infirmity.

Social Trends

A week rarely passes without some mention in the news of the obesity epidemic, both in adults and children. A gradual but persistent trend toward greater physical inactivity has emerged over the last half century. In lean budgetary times, governments and school districts are forced to make program cuts. Unfortunately, the first cuts often include reductions in or elimination of physical education and performing arts programs. The short-term benefit is a balanced budget; the long-term consequence is a population with a sedentary mind-set and a poor record of physical activity, health, and wellness. Our children, in general, spend more time involved in the on-screen action of a video game than they do being part of the action themselves. The combination of physical inactivity and poor nutrition is creating a time bomb, with monumental consequences at both the individual and societal levels. Hypoactivity is associated with many debilitating conditions such as cardiovascular disease, osteoporosis, and diabetes, and we will all pay the price, in the form of spiraling health care costs, workplace absence, and worker's compensation. For all too many, the result will be the personal tragedy of injury, disease, and a lower quality of life. The solution is undeniably complex but must include an increase in physical activity. Movement must be part of the solution.

Limits to Human Performance

Questions often arise concerning the limits of human performance. Is there a limit to our physical capabilities, and if so, what are those limits? In 1968, when Bob Beamon soared 8.90 m (more than 29 ft) to win the Olympic gold medal in the long jump, most everyone said the record would never be broken. The combination of Mexico City's mile-high altitude and a freakishly perfect jump, it was argued, produced a record thought to be unapproachable. And for almost 23 years, the record did go untouched. But as often happens, predictions prove untrue. In 1991, Mike Powell soared 8.95 m to break the record. Will Powell's record be broken? Most likely. But what is the ultimate limit? Clearly, no one will long jump as far as 15 m, so where between 8.95 m and 15 m is the limit? It is a tantalizing question with no certain answer.

Exceeding perceived limits to performance is not restricted to world-class levels. All too often, we regrettably hear the doctor of a paralyzed spinal cord injury patient predict, "She will never walk again." The determined patient, however, surpasses all expectations and learns to walk again. Clearly, the limits to performance depend on many often unexplainable factors. How do we estimate the effect of determination and inspiration on pushing the limits of human performance? Again, a good question with no certain answer.

Whether there are limits to performance or not, it is safe to say that we will never reach a limit to our interest in human movement. In many ways, movement will always hold some mystery for us. And as Albert Einstein said, "The most beautiful thing we can experience is the mysterious" (Einstein, 2005).

> It is a paradox to say the human body has no 'limit.' There must be a limit to the speed at which men can run. I feel this may be around 3:30 for the mile. However, another paradox remains—if an athlete manages to run 3:30, another runner could be found to marginally improve on that time.
>
> —Sir Roger Bannister, who ran the first sub-4-minute mile in 1954

critical thinking questions

1. What technological advances might occur in the next 20 years that would affect the study of human movement?

2. What future advances in medicine might enhance, or hinder, our ability to move?

3. What social and demographic trends, in addition to aging and physical inactivity, might affect the movement-related issues we will confront in the future?

4. Select three current world records and predict whether they will be broken in the next 10 years. Why have you made that prediction? What do you think are the limits to human performance for that record?

suggested readings

Aukstakalnis, S., & Blatner, D. (1992). *Silicon mirage: The art and science of virtual reality.* Berkeley, CA: Peachpit Press.

Benyus, J.M. (1997). *Biomimicry: Innovation inspired by nature.* New York: William Morrow.

Cetron, M., & Davies, O. (1997). *Probable tomorrows: How science and technology will transform our lives in the next twenty years.* New York: St. Martin's Press.

Negroponte, N.P. (1995). *Being digital.* New York: Knopf.

Spirduso, W.W., Francis, K.L., & MacRae, P.G. (2005). *Physical dimensions of aging* (2nd ed.). Champaign, IL: Human Kinetics.

appendix

Dynatomy Practice Problems

Now that you have seen how the muscle control formula is applied to a variety of simple movements, it is your turn to solve some analysis problems on your own. As emphasized earlier, the best way to learn how to use the formula is through practice. The charts on the following pages contain a broad selection of simple movements for your analysis. Please do not skip these practice problems; their mastery is essential in developing your skills as a student of human movement. The time you spend doing these exercises will be richly rewarded. With practice, you will soon be able to analyze the muscle actions involved in any movement task.

For each of the following problems, use the six steps of the muscle control formula to analyze the muscle actions at the specified joint. Movements identified as "slow" are performed at a speed slower than what the external force would produce in the absence of muscle action. Movements identified as "fast" are performed at a speed faster than what the external force would produce acting alone.

For solutions, see the solution table on pages 222 and 223.

Problem 1

a   b

Apply the muscle control formula for movement at the hip from position A to position B.

Step 1: _____

Step 2: _____

Step 3: _____

Step 4: _____

Step 5: _____

Step 6: _____

Apply the muscle control formula for movement at the hip from position B to position A (slow).

Step 1: _____

Step 2: _____

Step 3: _____

Step 4: _____

Step 5: _____

Step 6: _____

From *Dynatomy* by William C. Whiting and Stuart Rugg, 2006, Champaign, IL: Human Kinetics.

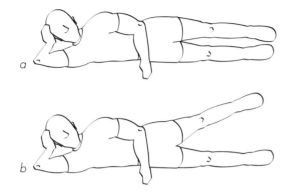

a

b

Apply the muscle control formula for movement at the hip from position A to position B.

Step 1: _____

Step 2: _____

Step 3: _____

Step 4: _____

Step 5: _____

Step 6: _____

Apply the muscle control formula for the leg held in position B.

Step 1: _____

Step 2: _____

Step 3: _____

Step 4: _____

Step 5: _____

Step 6: _____

Apply the muscle control formula for movement at the hip from position B to position A (slow).

Step 1: _____

Step 2: _____

Step 3: _____

Step 4: _____

Step 5: _____

Step 6: _____

a

b

Apply the muscle control formula for movement at the knee from position A to position B (slow).

Step 1: _____

Step 2: _____

Step 3: _____

Step 4: _____

Step 5: _____

Step 6: _____

Apply the muscle control formula for movement at the knee from position B to position A (slow).

Step 1: _____

Step 2: _____

Step 3: _____

Step 4: _____

Step 5: _____

Step 6: _____

From *Dynatomy* by William C. Whiting and Stuart Rugg, 2006, Champaign, IL: Human Kinetics.

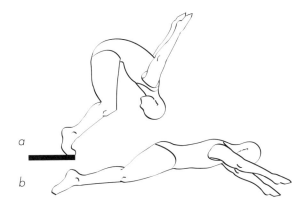

a

b

Apply the muscle control formula for movement at the hip from position A to position B.

Step 1: _____

Step 2: _____

Step 3: _____

Step 4: _____

Step 5: _____

Step 6: _____

From *Dynatomy* by William C. Whiting and Stuart Rugg, 2006, Champaign, IL: Human Kinetics.

Problem 5

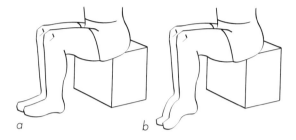

a   *b*

Apply the muscle control formula for the ankle held in anatomical position (position A).

Step 1: _____

Step 2: _____

Step 3: _____

Step 4: _____

Step 5: _____

Step 6: _____

Apply the muscle control formula for movement at the ankle from position A to position B (fast).

Step 1: _____

Step 2: _____

Step 3: _____

Step 4: _____

Step 5: _____

Step 6: _____

Apply the muscle control formula for movement at the ankle from position B to position A.

Step 1: _____

Step 2: _____

Step 3: _____

Step 4: _____

Step 5: _____

Step 6: _____

From *Dynatomy* by William C. Whiting and Stuart Rugg, 2006, Champaign, IL: Human Kinetics.

Problem 6

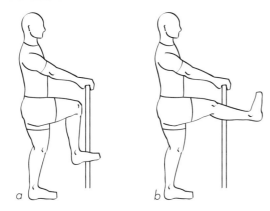

a   b

Apply the muscle control formula for movement at the knee from position A to position B.

Step 1: _____

Step 2: _____

Step 3: _____

Step 4: _____

Step 5: _____

Step 6: _____

Apply the muscle control formula for the leg held in position B.

Step 1: _____

Step 2: _____

Step 3: _____

Step 4: _____

Step 5: _____

Step 6: _____

Apply the muscle control formula for movement at the knee from position B to position A (slow).

Step 1: _____

Step 2: _____

Step 3: _____

Step 4: _____

Step 5: _____

Step 6: _____

a   *b*

Apply the muscle control formula for movement at the knee from position A to position B. (Note, the legs are pushing against resistance.)

Step 1: _____

Step 2: _____

Step 3: _____

Step 4: _____

Step 5: _____

Step 6: _____

Apply the muscle control formula for movement at the knee from position B to position A (slow).

Step 1: _____

Step 2: _____

Step 3: _____

Step 4: _____

Step 5: _____

Step 6: _____

From *Dynatomy* by William C. Whiting and Stuart Rugg, 2006, Champaign, IL: Human Kinetics.

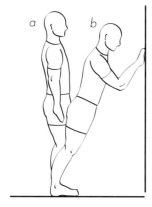

Apply the muscle control formula for movement at the ankle from position A to position B (slow).

Step 1: _____

Step 2: _____

Step 3: _____

Step 4: _____

Step 5: _____

Step 6: _____

Apply the muscle control formula for movement at the ankle from position B to position A (slow).

Step 1: _____

Step 2: _____

Step 3: _____

Step 4: _____

Step 5: _____

Step 6: _____

From *Dynatomy* by William C. Whiting and Stuart Rugg, 2006, Champaign, IL: Human Kinetics.

215

Apply the muscle control formula for movement at the shoulder (glenohumeral joint) from position A to position B.

Step 1: _____

Step 2: _____

Step 3: _____

Step 4: _____

Step 5: _____

Step 6: _____

Apply the muscle control formula for movement at the shoulder (glenohumeral joint) from position B to position A (slow).

Step 1: _____

Step 2: _____

Step 3: _____

Step 4: _____

Step 5: _____

Step 6: _____

From *Dynatomy* by William C. Whiting and Stuart Rugg, 2006, Champaign, IL: Human Kinetics.

Problem 10

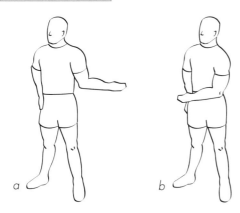

a   b

Apply the muscle control formula for movement at the left shoulder from position A to position B (slow).

Step 1: _____

Step 2: _____

Step 3: _____

Step 4: _____

Step 5: _____

Step 6: _____

Apply the muscle control formula for movement at the left shoulder from position B to position A (slow).

Step 1: _____

Step 2: _____

Step 3: _____

Step 4: _____

Step 5: _____

Step 6: _____

Problem 11

Apply the muscle control formula for the trunk held in the position shown.

Step 1: _____

Step 2: _____

Step 3: _____

Step 4: _____

Step 5: _____

Step 6: _____

Problem 12

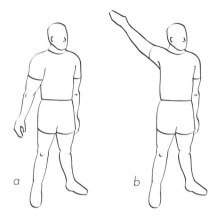

a   b

Apply the muscle control formula for movement at the shoulder from position A to position B.

Step 1: _____

Step 2: _____

Step 3: _____

Step 4: _____

Step 5: _____

Step 6: _____

Apply the muscle control formula for movement at the shoulder from position B to position A (fast).

Step 1: _____

Step 2: _____

Step 3: _____

Step 4: _____

Step 5: _____

Step 6: _____

a   b

Apply the muscle control formula for movement at the shoulder from position A to position B.

Step 1: _____

Step 2: _____

Step 3: _____

Step 4: _____

Step 5: _____

Step 6: _____

Apply the muscle control formula for movement at the shoulder from position B to position A (slow).

Step 1: _____

Step 2: _____

Step 3: _____

Step 4: _____

Step 5: _____

Step 6: _____

From *Dynatomy* by William C. Whiting and Stuart Rugg, 2006, Champaign, IL: Human Kinetics.

Problem 14

a   *b*

Apply the muscle control formula for movement at the elbow from position A to position B.

Step 1: _____

Step 2: _____

Step 3: _____

Step 4: _____

Step 5: _____

Step 6: _____

Apply the muscle control formula for the elbow held in position B.

Step 1: _____

Step 2: _____

Step 3: _____

Step 4: _____

Step 5: _____

Step 6: _____

Apply the muscle control formula for movement at the elbow from position B to position A (slow).

Step 1: _____

Step 2: _____

Step 3: _____

Step 4: _____

Step 5: _____

Step 6: _____

| Problem # and joint | Movement | [All movements begin from anatomical position] | | | | Notes |
		Position or movement	Movement plane	Type of muscle action	Active muscles (agonists)	
1. Hip	A-B	Adduction	Frontal	Concentric	Adductors (longus, brevis, magnus), pectineus, gracilis	
	B-A (slow)	Abduction	Frontal	Eccentric	[Same muscles as in (a)]	
2. Hip	A-B	Abduction	Frontal	Concentric	Gluteus medius, gluteus minimus, sartorius, tensor fascia lata, superior fibers of gluteus maximus	
	Hold	Abducted	Frontal	Isometric	[Same muscles as in (a)]	
	B-A	Adduction	Frontal	Eccentric	[Same muscles as in (a)]	
3. Knee	A-B (slow)	Extension	Sagittal	Concentric	Vastus medialis, vastus lateralis, vastus intermedius, rectus femoris	Both movements are "across gravity" and therefore the muscle action is concentric for both
	B-A (slow)	Flexion	Sagittal	Concentric	Biceps femoris, semitendinosus, semimembranosus	
4. Hip	A-B	Extension	Sagittal	Concentric	Gluteus maximus, long head of biceps femoris, semitendinosus, semimembranosus, posterior fibers of gluteus maximus	
5. Ankle	Hold	Neutral	Sagittal	Isometric	Tibialis anterior, peroneus tertius, extensor digitorum longus	
	A-B (fast)	Plantar flexion	Sagittal	Concentric	Gastrocnemius, soleus, peroneus longus & brevis, tibialis posterior	
	B-A	Dorsiflexion	Sagittal	Concentric	[Same muscles as in (a)]	
6. Knee	A-B	Extension	Sagittal	Concentric	Vastus medialis, vastus lateralis, vastus intermedius, rectus femoris	
	Hold	Extended	Sagittal	Isometric	[Same muscles as in (a)]	
	B-A (slow)	Flexion	Sagittal	Eccentric	[Same muscles as in (a)]	
7. Knee	A-B	Extension	Sagittal	Concentric	Vastus medialis, vastus lateralis, vastus intermedius, rectus femoris	
	B-A	Flexion	Sagittal	Eccentric	[Same muscles as in (a)]	

(continued)

Problem # and joint	Move-ment	Position or movement	Move-ment plane	Type of muscle action	Active muscles (agonists)	Notes
		[All movements begin from anatomical position]				
8. Ankle	A-B (slow)	Dorsiflexion	Sagittal	Eccentric	Gastrocnemius, soleus, peroneus longus & brevis, tibialis posterior, flexor hallucis longus, flexor digitorum longus, plantaris	
	B-A (slow)	Plantar flexion	Sagittal	Concentric	[Same muscles as in (a)]	
9. Shoulder	A-B	Flexion	Sagittal	Concentric	Anterior deltoid, pectoralis major (clavicular), coracobrachialis	The joint being assessed is the shoulder (glenohumeral) joint, not the elbow
	B-A (slow)	Extension	Sagittal	Eccentric	[Same muscles as in (a)]	
10. Shoulder	A-B (slow)	Internal rotation	Transverse	Concentric	Subscapularis, teres major, latissimus dorsi, pectoralis major, anterior deltoid	Both movements are "across gravity" and therefore the muscle action is concentric for both
	B-A (slow)	External rotation	Transverse	Concentric	Infraspinatus, teres minor, posterior deltoid	
11. Trunk	Hold	Flexed	Sagittal	Isometric	Rectus abdominis, internal obliques, external obliques	
12. Shoulder	A-B	Abduction	Frontal	Concentric	Supraspinatus, middle deltoid	(b) is concentric because the movement is fast and therefore involves the adductor muscles rather than the abductors, as would be the case if the movement were performed slowly
	B-A (fast)	Adduction	Frontal	Concentric	Pectoralis major, latissimus dorsi, teres major	
13. Shoulder	A-B	Horizontal adduction (flexion)	Transverse	Concentric	Anterior deltoid, pectoralis major, coracobrachialis, short head of biceps brachii	The joint being assessed is the shoulder (glenohumeral) joint, not the elbow
	B-A (slow)	Horizontal abduction (extension)	Transverse	Eccentric	[Same muscles as in (a)]	
14. Elbow	A-B	Flexion	Sagittal	Concentric	Biceps brachii, brachialis, brachioradialis	
	Hold	Flexed	Sagittal	Isometric	[Same muscles as in (a)]	
	B-A (slow)	Extension	Sagittal	Eccentric	[Same muscles as in (a)]	

glossary

abduction—Joint motion in the frontal plane (relative to anatomical position) that takes a segment away from the body's midline.

acetylcholine—Chemical neurotransmitter used by neuron terminal endings. The neurotransmitter used by all lower motor neurons that innervate skeletal muscles.

actin—Contractile protein that forms the backbone of the thin filaments within a sarcomere.

action—Internal state in which a muscle actively exerts a force, regardless of whether it shortens or lengthens. Also *contraction.*

action potential—Electrical signal that passes along the membrane of a neuron or muscle fiber. May also be called a nerve impulse with respect to a neuron or a muscle action potential as it relates to a muscle fiber.

active support—Contribution of muscle action to joint stability.

adaptive postural control—Strategy to maintain balance by modifying movement in response to situational changes.

adduction—Joint motion in the frontal plane (relative to anatomical position) that moves a segment toward the body's midline.

adenosine triphosphate—High-energy molecule used to supply energy for muscle contraction and other bodily functions.

adipocytes—Fat cells.

aerodynamic force—Force related to the motion of air and objects moving through the air.

agonist—Muscle actively involved in producing or controlling a movement.

amenorrhea—Absence of menstrual cycles.

amphiarthrosis—Functional classification of a joint with limited movement.

anatomical position—Erect posture with head facing forward and arms hanging straight down with palms facing forward.

anatomy—Study of the structure of organisms.

angular displacement—Angular measure from the starting position to the finishing position of an angular movement.

angular kinetic energy—(1/2) × (moment of inertia) × (angular velocity)^2

angular momentum—Product of a body's mass moment of inertia and its angular velocity.

angular motion—Motion in which a body rotates about an axis. Also *rotational motion.*

annulus fibrosus—Layered fibrocartilage network surrounding the nucleus pulposus in an intervertebral disc.

antagonist—Muscle acting in opposition to a movement.

anticipatory postural control—Strategy to maintain balance using anticipatory actions.

aponeuroses—Sheets of tendonlike material that cover a muscle's surface or connect a muscle to another muscle, or muscle to bone.

appendicular skeleton—Subsystem of the skeletal system containing bones of the pelvic and pectoral girdles and the limbs.

area moment of inertia—Measure of a body's resistance to bending.

arthritis—Joint inflammation.

arthroplasty—Joint replacement through surgical procedures.

articular cartilage—Smooth, shiny layer of hyaline cartilage covering the joint surfaces of articulating bones.

atrophy—To decrease in size.

axial skeleton—Subsystem of the skeletal system containing the skull, spinal column, and thoracic cage.

axis of rotation—Line (imaginary) about which joint rotation occurs.

balance—Maintenance of postural stability or equilibrium. Also *postural control.*

ballistics—Study of projectile motion.

base of support—Area within the outline of all contact points with the ground.

biarticular—Having action at two joints.

biomechanics—Application of mechanical principles to the study of biological organisms and systems.

body—Any collection of matter.

buoyant force—Equal and opposite force exerted by a liquid against the weight of a body, allowing the body to float.

bursa—Fluid-filled sac that helps cushion or reduce friction.

cadence—Gait tempo measured in steps per minute.

cancellous bone—Bone with high porosity (low density). Also *spongy* or *trabecular* bone.

cardiac muscle—Muscle tissue in the heart that is responsible for generating the forces that pump blood.

cartilage—Stiff connective tissue whose ground substance is nearly solid (of three types: hyaline, fibrocartilage, elastic).

cartilaginous joint—Structural classification of a joint bound by cartilage.

center of gravity—Point at which the weight of a body may be considered a concentrated mass without altering the motion of the body.

center of mass—Point about which the mass of a body is equally distributed.

cerebral palsy—Nonprogressive condition of muscle dysfunction and paralysis caused by brain injury at or near the time of birth.

circumduction—Special form of angular motion in which the distal end of a limb or segment moves in a circular pattern about a relatively fixed proximal end, tracing out a cone-shaped pattern.

coactivation—Simultaneous action of agonists and antagonists at a given joint.

compact bone—Bone with high density (low porosity). Also *cortical bone.*

compliance—Measure of the relationship between strain and stress. Inverse of stiffness.

compression—Action tending to push together.

concentric—Shortening muscle contraction. The torque produced by the muscle is greater than the external torque, and therefore, the muscle is able to shorten while overcoming the external load.

connective tissue—Class of tissue that provides support and protection and binds tissues together (e.g., bone, tendons, ligaments, cartilage, adipose, blood).

conservation (momentum, energy)—No net gain or loss (of momentum or energy).

contractile component—Structure within a muscle that can produce force. For example, the fundamental contractile component of a skeletal muscle is the sarcomere.

contractility—A muscle's ability to generate a pulling, or tension, force.

contraction—Internal state in which a muscle actively exerts a force, regardless of whether it shortens or lengthens. Also *action.*

coordination—Muscles working together with correct timing and intensity to produce or control a movement.

cortex—Hard outer covering of a bone. Also *cortical shell.*

cortical bone—Bone with high density (low porosity). Also *compact bone.*

cortical shell—Protective outer bony surface. Also *cortex.*

crawling—Form of infant locomotion with progression on hands and stomach.

creeping—Form of infant locomotion with movement on hands and knees.

curvilinear motion—Motion along a curved line.

deformational energy—Energy stored when a body is deformed. Also *strain energy.*

delayed onset muscle soreness—The muscle soreness that may occur 24 to 48 hours after an exercise session.

demi plié—dance movement involving half bending of the knees

density (bone)—Measure of the hard tissue in bone (hydroxyapatite crystal per unit volume); inverse of porosity.

depression—Movement of a structure in an inferior, or downward, direction.

diaphysis—Shaft of a long bone.

diarthrosis—Functional classification of a freely movable joint.

digitization—Process used to quantitatively identify the location (coordinates) of anatomical landmarks using computerized analysis systems.

distance—Scalar measure of how far a body has moved.

dorsiflexion—Joint motion at the ankle where the foot moves toward the lower leg.

double support—Period when the body's weight is supported by both legs.

drag force—Component of force that acts parallel to the direction of an object's motion in a fluid.

dynamic equilibrium—State in which the accelerations are balanced according to Newton's second law of motion (i.e., force = mass × acceleration).

dynamic posture—Posture of motion (as seen in walking and running).

dynatomy—Study of human anatomy from a dynamic, or movement-focused, perspective.

eccentric—Lengthening muscle contraction. The torque produced by the muscle is less than the external torque, but the torque produced by the muscle causes the joint movement to occur more slowly than the external torque would tend to make the limb move.

economy—Amount of metabolic energy required to perform a given amount of work.

edema—An abnormal accumulation of fluid within a structure, organ, or tissue.

efficiency—Amount of mechanical output produced for a given amount of metabolic input (i.e., how much work can be done using a given amount of energy).

elastic cartilage—Flexible cartilage found in areas where extensibility is needed (e.g., external ear and respiratory system).

elasticity—A tissue's ability to return to its original length and shape after an applied force is removed.

electromyography—Study of the electrical activity of muscles.

elevation—Movement of a structure in a superior, or upward, direction.

endomysium—Fibrous connective tissue sheath that surrounds each muscle fiber located within a fascicle in a skeletal muscle.

energy (mechanical)—Ability or capacity to do mechanical work.

epimysium—Fibrous connective tissue sheath that surrounds a whole skeletal muscle.

epiphyseal growth plate—Region of hyaline cartilage found between the diaphysis and epiphysis in developing bone.

epiphysis—An end of a long bone (plural is epiphyses**).**

equilibrium—State of balance between opposing forces or actions.

equinus—Foot position in which the forefoot is lower than the heel. In gait, at initial contact, the foot is plantar flexed.

eversion—Joint motion at the intertarsal joints that results in the sole of the foot being moved away from the body's midline.

excitability—Describes the ability of muscle to respond to a stimulus. Also *irritability.*

excitation–contraction coupling—Sequence of events involved in producing a muscle contraction, from exocytosis of the lower motor neuron through the interaction between actin and myosin filaments.

exocytosis—Process of fusion of a vesicle to the presynaptic membrane and the subsequent release of its neurotransmitter.

extensibility—Describes the ability of muscle to lengthen, or stretch, and as a consequence, to generate force over a range of lengths.

extension—Joint motion in the sagittal plane (relative to anatomical position) in which the angle between articulating segments increases.

external rotation—Joint motion in the transverse plane (relative to anatomical position) that rotates a segment's anterior surface away from the body's midline. Also *lateral rotation.*

extracellular matrix—Noncellular component of a tissue.

fascicle—Bundle or collection of fibers, usually of muscle or nerve fibers.

fibrocartilage—Strong and flexible cartilage that reinforces stress points and serves as filler material in and around joints.

fibrous joint—Structural classification of a joint bound by connective tissues composed primarily of collagen fibers.

fibrous joint capsule—Fibrous encasement surrounding a synovial joint.

fine motor skills—Skills involving intricate movements of small joints.

first-class lever—Lever system with the axis between the resistance force and the effort force.

flexion—Joint motion in the sagittal plane (relative to anatomical position) in which the angle between articulating segments decreases.

flight phase—Period of the running gait cycle when both feet are off the ground (i.e., period of nonsupport).

float phase—See *flight phase.*

fluid mechanics—Branch of mechanics dealing with the properties and behavior of gases and liquids.

force—Mechanical action or effect applied to a body that tends to produce acceleration.

force–velocity relationship—Property of skeletal muscle that shows its force production capability is dependent on its contraction velocity.

form drag—Resistive force produced when an object parts the medium through which it is passing.

friction—Resistance developed at the interface of two surfaces, acting opposite the direction of motion or impending motion.

frontal plane—Plane dividing the body into anterior and posterior sections.

gait—Form of locomotion, usually referring to walking or running.

gait cycle—Sequential occurrence of a stance and swing phase for a single limb.

general motion—Combined linear and angular motion.

glycolytic—The ability to metabolize glucose for energy.

gouty arthritis (gout)—Joint inflammation caused by uric acid crystals embedded in joint structures, leading to irritation and inflammation.

grand plié—dance movement involving full bending of the knees.

gravitational potential energy—Energy possessed by a body as a result of its position.

gross motor skills—Skills involving movement and control of the limbs.

ground reaction force—Equal and opposite force exerted by the ground against an object contacting it.

ground substance—Nonfibrous component of the extracellular matrix.

growth plate—See *epiphyseal growth plate.*

hematopoiesis—Process of blood cell production.

histology—Study of tissue structure.

hyaline cartilage—Cartilage found on joint surfaces, on anterior surfaces of the ribs, and in areas of the respiratory system. Also serves as the precursor to bone in the developing fetus.

hyperextension—Joint motion in the sagittal plane (relative to anatomical position) in which the angle between articulating segments increases beyond anatomical position.

hypertrophy—To increase in size.

idealized force vector—Single vector used to represent many vectors.

impingement syndrome—Pathological condition in which pressure increases within a confined anatomical space and the enclosed tissues are detrimentally affected.

inertia—Resistance to a change in a body's state of linear motion.

initial contact—First foot contact with the ground to end the swing phase and begin the stance phase.

innervation—Connection between a nerve fiber (axon including the terminal endings) and another structure such as a muscle fiber.

insertion—Site of musculotendinous attachment on the more movable end of a bone.

intercellular substance—Noncellular component of a tissue.

internal rotation—Joint motion in the transverse plane (relative to anatomical position) that rotates a segment's anterior surface inward toward the body's midline. Also *medial rotation.*

interosseous membrane—Collagenous tissue binding two bones together (e.g., tibiofibular joint).

intervertebral disc—Structure between two adjacent vertebrae composed of an inner gelatinous mass (nucleus pulposus) and a surrounding layered fibrocartilage network (annulus fibrosus).

inversion—Joint motion at the intertarsal joints that results in the sole of the foot being moved inward toward the body's midline.

irritability—Describes the ability of muscle to respond to a stimulus. Also *excitability.*

isoinertial—Represents constant resistance. A more accurate term than isotonic for human muscle contractions.

isokinetic—Describes a contraction performed with a constant angular velocity.

isometric—Refers to constant length of the musculotendinous unit and hence no limb movement. Torque produced by the muscle is equal and opposite to the external torque.

isotonic—Literally means constant tension. This condition does not occur in intact human subjects (i.e., *in vivo*) because the level of muscle force varies continuously and rarely, if ever, is constant throughout a movement.

kinematics—Description of motion with respect to space and time, without regard to the forces involved.

kinesiology—Study of the art and science of human movement.

kinetic energy—Energy possessed by a body by virtue of its motion.

kinetic friction—Frictional resistance created while an object is moving (i.e., sliding, rolling) along a surface.

kinetics—Assessment of motion with regard to forces and force-related measures.

knee extensor mechanism—Functional unit, made up of the quadriceps muscles, patella, and ligament attachment to the tibia, that functions to extend the knee.

kyphosis—Sagittal plane spinal deformity characterized by excessive flexion, usually in the thoracic region.

labrum—U-shaped ring of fibrocartilage around the rim of a joint.

lacuna—Small pocket or space.

laminar flow—Flow characterized by a smooth, parallel pattern of fluid motion.

lateral flexion—Sideways bending of the vertebral column in the frontal plane.

lateral rotation—See *external rotation*.

length–tension relationship—Property of skeletal muscle that shows its force production capability is dependent on the length of the muscle's contractile and noncontractile structures.

lever—Rigid structure, fixed at a single point, to which two forces are applied at two different points.

lever arm—See *moment arm*.

lift force—Component of force that acts perpendicular to the direction of an object's motion in a fluid.

ligament—Connective tissue that connects bone to bone.

line of force action—Line along which a force acts, extending infinitely in both directions along the line of a finite force vector.

linear displacement—Straight-line vector measure from the starting point to the finishing point of a movement.

linear kinetic energy—(1/2) × (mass) × (linear velocity)^2.

linear momentum—Product of a body's mass and its velocity.

linear motion—Motion along a straight or curved line.

load—Force applied externally to a body.

locomotion—Process of moving from one place to another place.

lordosis—Sagittal plane spinal deformity characterized by excessive extension, usually in the lumbar region.

lower motor neuron—Nerve cell that originates in the central nervous system and innervates a skeletal muscle.

magnus force—Force created by the spin of an object that creates a deviation in its normal trajectory.

marrow—Loose connective tissue found in the cavities and spaces in bone.

mass—Amount of matter, or substance, constituting a body.

mass moment of inertia—Measure of a body's resistance to rotation about an axis.

material mechanics—Branch of mechanics dealing with the internal response of materials (tissues) to external loads.

medial epicondylitis—Inflammation at the medial epicondyle of the humerus, often due to excessive throwing.

medial rotation—See *internal rotation*.

median plane—Sagittal plane that divides the body in half. Also *midsagittal plane.*

meniscus—Fibrocartilage pads interposed between bones to provide shock absorption and improve bony fit.

mesenchymal cells—Unspecialized cells that later differentiate into specific cell types.

midsagittal plane—Sagittal plane that divides the body in half. Also *median plane.*

mobility (joint)—Ability of a joint to move through a range of motion.

mobility (movement)—Ability to move readily.

modeling—Formation of new bone.

moment—Effect of a force that tends to cause rotation about an axis.

moment arm—Perpendicular distance from the axis of rotation to the line of force action. Also *torque arm* and *lever arm.*

moment of inertia—Resistance to a change in a body's state of angular motion.

momentum—Property of a moving body that is determined by the product of the body's mass and its velocity.

motion segment—Two adjacent vertebrae and the intervening intervertebral disc.

motor—Involving muscular movement.

motor behavior—Study of the behavioral aspects of movements, including development, learning, and control.

motor control—Study of the neural, physical, and behavioral aspects of movement.

motor development—Study of changes in movement behavior throughout the life span.

motor learning—Study of how motor skills are learned.

motor milestones—Landmark events in the progression of learning muscular control of movement.

motor skill—Voluntary movement used to complete a desired task action or achieve a specific goal.

motor unit—A single lower motor neuron plus all the muscle fibers it innervates.

multiarticular—Having action at more than three joints.

muscle control formula—Step-by-step procedure for determining the involved muscles and their action for any joint movement.

muscle redundancy—Situation in which more muscles are available to perform an action than are minimally necessary.

muscle synergy—Cooperative action of several muscles working together as a single unit.

musculotendinous junction—Region where a muscle and tendon connect. Also *myotendinous junction*.

musculotendinous unit—Combined unit including a skeletal muscle and its tendons.

myomesin—Structural protein that forms the M line within a sarcomere and helps maintain the alignment of the myosin filaments.

myosin—Contractile protein that forms the thick filaments within a sarcomere.

myotendinous junction—Region where a muscle and tendon connect. Also *musculotendinous junction*.

net moment—Sum of all the moments (torques) acting on a body. Also *net torque*.

net torque—See *net moment*.

neutral position—Reference position for joints in anatomical position.

neutralization—Canceling, or neutralizing, action of two or more muscles.

nilotic stance—Posture in which a person stands on one leg with the opposite leg used to brace the standing knee (i.e., like a flamingo).

nomenclature—A generally agreed upon system or set of terms.

noncontractile component—Structure within a muscle that, by itself, cannot produce force (e.g., the connective tissue sheaths within a skeletal muscle).

normal force—Component of force acting on an object, perpendicular to the surface.

nucleus pulposus—Gelatinous mass on the inside of an intervertebral disc.

oligomenorrhea—Irregular menstrual cycles.

opposition—Ability of the thumb to work with the other four fingers to perform grasping movements.

organ—Structure composed of two or more tissues that have a definite form and function.

organ system—Group of organs that work together to perform specific functions.

origin—Site of musculotendinous attachment on the less movable end of a bone.

osteoarthritis—Degradation of articular cartilage caused by mechanical action. The most common type of arthritis.

osteoblast—Mononuclear bone cells that produce new bone material.

osteoclast—Large, multinucleated cells that break down, or resorb, bone.

osteocyte—Mature bone cells that are smaller and less active than osteoblasts.

osteoid—Organic portion of the extracellular matrix in bone.

osteon—Fundamental structural unit of compact bone.

osteopenia—Mild to moderate bone loss.

osteoporosis—Severe bone loss that increases the risk of fracture.

osteoprogenitor cells—Undifferentiated mesenchymal cells with the ability to produce daughter cells that can differentiate to become osteoblasts.

osteotendinous junction—Region where a bone and tendon connect.

oxidative—The ability to utilize oxygen for aerobic metabolism.

paddling—Stroke technique where the swimmer uses a straight pull with the hand oriented 90° to the surface of the water.

palpation—To examine by touch.

parasagittal plane—Any sagittal plane offset from the midline to one side or the other.

passive support—Contribution of noncontractile tissues to joint stability.

patellar tracking—Sliding movement of the patella in the intercondylar groove of the femur.

peak bone mass—Highest amount of bone mass during one's lifetime.

pectoral girdle—Bony complex made up of the clavicle and scapula.

pedaling cycle—One revolution of the pedal starting from, and returning to, the top dead center position of the crank arm.

pelvic girdle—Bony complex made up of the ilium, ischium, and pubis.

pennation—Angle formed between the muscle's line of pull and the orientation of the muscle fibers.

perception–action coupling—Response of an infant to environmental cues.

perimysium—Fibrous connective tissue sheath that surrounds each fascicle within a skeletal muscle.

periosteum—Fibrous connective tissue covering bone except at the joint surfaces.

perturbation—Disturbance of motion.

physiology—Study of the function of body parts and systems.

pirouette—Complete turn of the body on one foot.

plantar flexion—Joint motion at the ankle where the foot moves away from the lower leg.

plasticity—Ability to adapt to environmental stimuli and events.

plié—Literally translated it means bent or bending. A bending of the knee or knees.

plyometrics—Form of training that consists of stretch–shorten cycles. In other words, cycles of eccentric and concentric contractions, such as box jumps.

point mass—Concentrated mass at a single point.

polar moment of inertia—Measure of a body's resistance to twisting, or torsion.

porosity (bone)—Measure of the soft tissue in bone; inverse of density.

postural control—Maintenance of postural stability, or equilibrium. Also *balance*.

postural sway—Small movement fluctuations used to maintain posture.

posture—Alignment or position of the body and its parts.

potential energy—Energy created by a body's position or deformation.

power—Rate of work production.

primary curvature (spine)—Spinal curvature first evident in the fetus and maintained after birth by the thoracic and sacral regions of the spine.

prime mover—Muscle most responsible for producing or controlling a given joint movement.

projectile—Object moving through space subject only to the effects of gravity and air resistance.

projection (center of gravity)—Line drawn from the center of gravity vertically downward to the base of support.

pronation (foot and ankle)—Combination of ankle dorsiflexion, subtalar eversion, and external rotation of the foot.

pronation (forearm)—Rotation of the radius over the ulna (from anatomical position).

prone—Lying posture with face pointed down.

protraction—Movement of a structure anteriorly, or toward the front of the body.

quadriceps avoidance—Gait strategy adopted by persons with ACL deficiency to avoid activation of muscles in the quadriceps group.

radial deviation—Abduction of the wrist and hand, such that the thumb moves closer to the radius.

range of motion—Measure of joint mobility.

reactive postural control—Strategy to maintain balance (e.g., to prevent a fall) using muscle action.

rectilinear motion—Motion along a straight line.

recumbent—Lying posture.

relevé—Rising on the toes by plantar flexing the ankle.

remodeling—Adaptation of existing bone through a process of resorption and replacement.

resistance training—Contracting muscles against resistance to enhance muscle strength, power, endurance, or size.

resorption—Breakdown or demineralization of bone.

retraction—Movement of a structure posteriorly, or toward the back of the body.

rheumatoid arthritis—Autoimmune condition leading to joint inflammation.

rotational motion—Motion in which a body rotates about an axis. Also *angular motion*.

running—Form of upright locomotion characterized by alternating periods of single support and a flight phase.

sagittal plane—Plane dividing the body into left and right sections.

sarcolemma—Membrane covering a skeletal muscle cell (fiber).

sarcomere—Contractile unit within skeletal muscle that spans from one Z disc to the next Z disc.

sarcoplasmic reticulum—Network of tubes and sacs (known as terminal cisternae) that surrounds each myofibril within a muscle fiber. The terminal cisternae store the calcium needed for muscle contraction.

scapulohumeral rhythm—Coordinated action of the humerus and scapula in facilitating glenohumeral abduction.

scoliosis—Lateral (frontal plane) curvature of the spine.

screw-home mechanism—Tibiofemoral rotation during the final few degrees of knee extension.

sculling—Stroke technique where the swimmer uses a curvilinear pull while varying the hand angle (angle of attack).

secondary curvature (spine)—Spinal curvature developed primarily after birth in the cervical and lumbar regions of the spine.

second-class lever—Lever system with the resistance force between the axis and the effort force.

sesamoid bone—Bones embedded in a tendon. Sometimes referred to as "floating" bones (e.g., patella).

shear—Action that tends to produce horizontal sliding of one layer over another.

single support—Period when the body's weight is supported by a single leg.

skeletal muscle—Muscle tissue responsible for maintaining posture and producing movement.

sliding filament model—Model used to describe how a sarcomere produces force. The final steps associated with excitation–contraction coupling that describe the interaction between the actin and myosin filaments needed to produce force.

smooth muscle—Muscle tissue that facilitates substance movement through tracts in the circulatory, respiratory, digestive, urinary, and reproductive systems.

spastic diplegia—Pathological gait characterized by abnormally flexed, adducted, and internally rotated hips; hyperflexed knees; and equinus of the foot and ankle.

spongy bone—Bone with high porosity (low density). Also *trabecular* or *cancellous* bone.

sprain—Injury to a musculotendinous unit.

stability (joint)—Ability of a joint to resist dislocation.

stability (movement)—Ability to resist movement.

stabilization—Muscle action to maintain, or stabilize, a position.

stance phase—Period during which the foot is in contact with the ground during gait.

static equilibrium—State of balance in which there is no net acceleration. Usually refers to stationary, or nonmoving, bodies.

static friction—Frictional resistance created while an object is not moving.

static postural control—Strategy to maintain a static posture, with the body's center of gravity projection kept within the base of support.

static posture—Postures involving little or no movement.

steady-state posture—Postures involving slight movement or swaying.

stem cells—See *mesenchymal cells*.

step—Period from initial contact of one leg to initial contact of the opposite leg.

step time—Duration of a single step; inverse of cadence.

stiffness—Measure of the relationship between stress and strain (i.e., how much a body deforms in response to a given load).

strain (injury)—Musculotendinous injury typically produced when too much force is transmitted through the musculotendinous unit.

strain (mechanical)—Deformation or change in length and shape of a tissue.

strain energy—Energy stored when a body is deformed. Also *deformational energy*.

strain-rate dependent—Property of a tissue dictating that its mechanical response to loading depends on the rate at which the tissue is deformed.

stress—Internal resistance developed in response to an externally applied force (load).

stretch–shorten cycle—Eccentric contraction immediately followed by a concentric contraction.

stride—Period from initial contact of one leg to initial contact of the same leg (one stride equals two steps).

subluxation—Partial joint dislocation.

supination (foot and ankle)—Combination of ankle plantar flexion, subtalar inversion, and internal rotation of the foot.

supination (forearm)—Rotation of the radius over the ulna (back to anatomical position).

supine—Lying posture with face pointed up.

surface drag—Resistive force produced directly over the surface of the object as it passes through a fluid medium.

suture joint—Fibrous joint connecting interlocking bones of the skull.

swing phase—Period during which the foot is not in contact with the ground during gait.

symphysis—Joint between two bones separated by a fibrocartilage pad.

synapse—Junction between a neuron and its target structure (e.g., another neuron, sarcolemma of skeletal muscle fiber). The region where a nervous impulse is passed from one neuron to another structure.

synarthrosis—Functional classification of a joint with no movement.

synchondrosis—Joint bound by hyaline cartilage.

syndesmosis—Joint bound by ligaments.

synovial fluid—Viscous fluid found in synovial joints that provides lubrication and reduces friction.

synovial joint—Structural classification of a joint containing a fibrous joint capsule, synovial membrane, synovial cavity, synovial fluid, and articular cartilage.

synovial joint cavity—Space between the bones in a synovial joint.

synovial membrane—Thin membrane, on the inner surface of the fibrous joint capsule of a synovial joint, that produces synovial fluid.

takeoff—Last foot contact with the ground to end the stance phase and begin the swing phase. Also *toe-off*.

tangential force—Component of force acting parallel to a body's surface.

temporal analysis—Mechanical assessment based on time duration.

tendinitis—Inflammation of a tendon, usually due to overuse.

tendon—Cordlike connective tissue that connects muscle to bone.

tension—Action tending to pull apart.

third-class lever—Lever system with the effort force between the axis and the resistance force.

tissue—Group of cells with similar function and their surrounding noncellular material.

titin—Structural protein that extends from the Z disc to both the myosin filament and the M line.

toe-off—Last foot contact with the ground to end the stance phase and begin the swing phase. Also *takeoff*.

torque—Effect of a force that tends to cause twisting, or torsion, about an axis. Also used as a synonym for *moment* as the effect of a force that tends to cause rotation of a body about an axis of rotation.

torque arm—See *moment arm*.

total mechanical energy—Sum of a body's linear kinetic energy, angular kinetic energy, and positional potential energy.

trabecular bone—Bone with high porosity (low density). Also *spongy* or *cancellous* bone.

trajectory—Path along which a projectile travels.

transfer (momentum, energy)—Exchange of momentum or energy from one body to another.

transverse plane—Plane dividing the body into superior and inferior sections.

transverse tubules (T-tubules)—Invaginations of the sarcolemma that pass through the muscle fiber between the myofibrils.

Trendelenberg gait—Pathological gait characterized by pelvic drop during early to midstance due to weakness or paralysis of the hip abductors.

triarticular—Having action at three joints.

tropomyosin—Regulatory protein located on the actin filament. In the relaxed skeletal muscle, it covers the myosin-head binding sites on the actin molecules.

troponin—Regulatory protein that binds to both tropomyosin and actin. When combined with calcium, troponin influences tropomyosin to initiate muscle contraction.

turbulent flow—Flow characterized by a chaotic pattern of fluid motion.

ulnar deviation—Adduction of the wrist and hand, such that the little (fifth) finger moves closer to the ulna.

uniarticular—Having action at one joint.

viscoelasticity—Describes a tissue's response to loading that is both strain-rate dependent and elastic.

viscosity—Resistance to flow.

walking—Form of upright locomotion in which at least one foot is always in contact with the ground.

wave drag—Resistive force produced when an object is moving near or along the surface of a fluid medium, usually water, causing waves to form.

weight—A measure of the effect of gravity on a mass.

work—Mechanical measure of force multiplied by displacement.

zygapophysis—Articular process between adjacent vertebrae.

references

Alter, M.J. (1996). *Science of flexibility* (2nd ed.). Champaign, IL: Human Kinetics.

Aukstakalnis, S., & Blatner, D. (1992). *Silicon mirage: The art and science of virtual reality.* Berkeley, CA: Peachpit Press.

Bartlett, J. (1968). *Familiar quotations* (14th ed.). Boston: Little, Brown and Company.

Bartlett, R. (2000). Principles of throwing. In V.M. Zatsiorsky (Ed.), *Biomechanics in sport: Performance enhancement and injury prevention* (pp. 365-380). Oxford, UK: Blackwell Science.

Basmajian, J.V., & DeLuca, C. (1985). *Muscles alive* (5th ed.). Baltimore: Williams & Wilkins.

Bergman, R.A., Thompson, S.A., Afifi, A.K., & Saadeh, F.A. (1988). *Compendium of human anatomic variation.* Baltimore: Urban & Schwarzenberg.

Bohannon, R.W. (1997). Comfortable and maximum walking speed of adults aged 20-79 years: Reference values and determinants. *Age and Ageing, 26,*15-19.

Branch, T., Partin, C., Chamberland, P., Emeterio, E., & Sabetelle, M. (1992). Spontaneous fractures of the humerus during pitching: A series of 12 cases. *American Journal of Sports Medicine, 20(4),* 468-470.

Burstein, A.H., & Wright, T.M. (1994). *Fundamentals of orthopaedic biomechanics.* Baltimore: Williams & Wilkins.

Cappozzo, A., & Marchetti, M. (1992). Borelli's heritage. In A. Cappozzo, M. Marchetti, & V. Tosi (Eds.), *Biolocomotion: A century of research using moving pictures* (pp. 33-47). Rome: Promograph.

Cech, D.J, & Martin, S.M. (2002). *Functional movement development across the life span.* Philadelphia: Saunders.

Colwin, C.M. (2002). *Breakthrough swimming.* Champaign, IL: Human Kinetics.

Cunningham, A. (Ed.) (2002). *Guinness World Records 2002.* London: Guinness World Records Ltd.

Cutler, W.B., Friedmann, E., & Genovese-Stone, E. (1993). Prevalence of kyphosis in a healthy sample of pre- and postmenopausal women. *American Journal of Physical Medicine and Rehabilitation, 72(4),* 219-225.

D'Israeli, B. (2005). The quotations page. Available: http://www.quotationspage.com/quote/26860.html [March 4, 2005].

Darwin, C. (1998). *The expression of the emotions in man and animals* (3rd ed.). New York: Oxford University Press.

Descartes, R. (1998). Translated by Donald A. Cress. *Discourse on method and meditations on first philosophy* (4th ed.). Indianapolis, IN: Hackett.

Einstein, A. (2005). The quotations page. Available: www.quotationspage.com/quote/1388.html [March 4, 2005].

Emerson, R.W. (2005). BrainyQuote. Available: http://brainyquote.com/quotes/quotes/r/ralphwaldo122015.html [March 4, 2005]

Enoka, R.M. (1994). *Neuromechanical basis of kinesiology* (2nd ed.). Champaign, IL: Human Kinetics.

Enoka, R.M. (2002). *Neuromechanics of human movement* (3rd ed.). Champaign, IL: Human Kinetics.

Fiatarone, M.A., Marks, E.C., Ryan, N.D., Meredith, C.N., Lipsitz, L.A., & Evans, W.J. (1990). High-intensity strength training in nonagenarians: Effects on skeletal muscle. *Journal of the American Medical Association, 263(22),* 3029-3034.

Fitts, P.M. (1964). Categories of human learning. In A.W. Melton (Ed.), *Perceptual-motor skills learning* (pp. 243-285). New York: Academic Press.

Gowan, I.D., Jobe, F.W., Tibone, J.E., Perry, J., & Moynes, D.R. (1987). A comparative electromyographic analysis of the shoulder during pitching: Professional versus amateur pitchers. *American Journal of Sports Medicine, 15(6),* 586-590.

Harrison, A.J., & Gaffney, S. (2001). Motor development and gender effects on stretch-shortening cycle performance. *Journal of Science and Medicine in Sport, 4(4),* 406-415.

Haywood, K.M., & Getchell, N. (2001). *Life span motor development* (3rd ed.). Champaign, IL: Human Kinetics.

Hudson, J.L. (1986). Coordination of segments in the vertical jump. *Medicine & Science in Sports & Exercise, 18(2),* 242-251.

Inman, V.T., Ralston, H.J., & Todd, F. (1994). Human locomotion. In J. Rose, & J.G. Gamble (Eds.), *Human walking* (pp. 1-22). Baltimore: Williams & Wilkins.

James, W. (1983). *Talks to teachers on psychology.* Cambridge, MA: Harvard University Press.

Jerome, J. (1980). *The sweet spot in time.* New York: Summit.

Jobe, F.W., Moynes, D.R., Tibone, J.E., & Perry, J. (1984). An EMG analysis of the shoulder in pitching: A second report. *American Journal of Sports Medicine, 12(3),* 218-220.

Judge, J.O., Ounpuu, S., & Davis, R.B. (1996). Effects of age on the biomechanics and physiology of gait. *Clinics in Geriatric Medicine, 12,* 659-678.

Keele, K.D. (1983). *Leonardo da Vinci's elements of the science of man*. New York: Academic Press.

Langendorfer, S.J., & Roberton, M.A. (2002). Individual pathways in the development of forceful throwing. *Research Quarterly in Exercise and Sport, 73(3)*, 245-256.

Levangie, P.K., & Norkin, C.C. (2001). *Joint structure and function: A comprehensive analysis* (3rd ed.). Philadelphia: Davis.

Lieber, R.L. (2002). *Skeletal muscle structure, function, and plasticity: The physiological basis of rehabilitation* (2nd ed.). Philadelphia: Lippincott Williams & Wilkins.

Luhtanen, P., & Komi, R.V. (1978). Segmental contribution to forces in vertical jump. *European Journal of Applied Physiology and Occupational Physiology, 38(3)*, 181-188.

Maki, B.E., & McIlroy, W.E. (1996). Postural control in the older adult. *Clinics in Geriatric Medicine, 12(4)*, 635-658.

Marcus, R., Cann, C., Madvig, P., Minkoff, U., Goddard, M., Bayer, M., Martin, M., Haskell, W., & Genant, H. (1985). Menstrual function and bone mass in elite women distance runners: Endocrine and metabolic features. *Annals of Internal Medicine, 102*, 158-163.

McGill, S. (2002). *Low back disorders: Evidence-based prevention and rehabilitation*. Champaign, IL: Human Kinetics.

Mero, A., & Komi, P.V. (1986). Force-, EMG-, and elasticity-velocity relationships at submaximal, maximal and supramaximal running speeds in sprinters. *European Journal of Applied Physiology, 55(5)*, 553-561.

Mish, F.C. (Ed.). (1997). *Merriam-Webster's collegiate dictionary* (10th ed.). Springfield, MA: Merriam-Webster.

Murray, M.P., Guten, G.N., Mollinger, L.A., & Gardner, G.M. (1983). Kinematic and electromyographic patterns of Olympic race walkers. *American Journal of Sports Medicine, 11(2)*, 68-74.

Neumann, D.A. (2002). *Kinesiology of the musculoskeletal system: Foundations for physical rehabilitation*. St. Louis: Mosby.

Newell, K.M. (1986). Constraints on the development of coordination. In M.G. Wade & H.T.A. Whiting (Eds.), *Motor development in children: Aspects of coordination and control* (pp. 341-361). Amsterdam: Martin Nijhoff.

O'Donoghue, D.H. (1984). *Treatment of injuries to athletes* (4th ed.). Philadelphia: Saunders.

O'Malley, C.D., & Saunders, J.B. de C.M. (1997). *Leonardo da Vinci on the human body*. Avenel, NJ: Wings Books.

Ostrosky, K.M., VanSwearingen, J.M., Burdett, R.G., & Gee, Z. (1994). A comparison of gait characteristics in young and old subjects. *Physical Therapy, 74(7)*, 637-646.

Padman, R. (1995). Scoliosis and spine deformities. *Delaware Medical Journal, 67(10)*, 528-533.

Pascal, B., & Krailsheimer, A.J. (1995). *Pensees*. New York: Penguin.

Perry, J. (1992). *Gait analysis: Normal and pathological function*. Thorofare, NJ: Slack, Inc.

Puniello, M.S., McGibbon, C.A., & Krebs, D.E. (2001). Lifting strategy and stability in strength-impaired elders. *Spine, 26(7)*, 731-737.

Roberton, M.A., & Halverson, L.E. (1984). *Developing children: Their changing movement*. Philadelphia: Lea & Febiger.

Robertson, D. & Mosher, R. (1985). Work and power of the leg muscles in soccer kicking. In D. Winter (Ed.), *Biomechanics IX-B* (pp. 533-538). Champaign, IL: Human Kinetics.

Schmidt, R.A., & Lee, T.D. (1998). *Motor control and learning: A behavioral emphasis*. Champaign, IL: Human Kinetics.

Schmitz, C., Martin, N., & Assaiante, C. (1999). Development of anticipatory postural adjustments in a bimanual load-lifting task in children. *Experimental Brain Research, 126(2)*, 200-204.

Sisto, D.J., Jobe, F.W., Moynes, D.R., & Antonelli, D.J. (1987). An electromyographic analysis of the elbow in pitching. *American Journal of Sports Medicine, 15(3)*, 260-263.

Smith, L.K., Weiss, E.L., & Lehmkuhl, L.D. (1996). *Brunnstrom's clinical kinesiology*. Philadelphia: Davis.

Stanislavski, C. (1948). *An actor prepares*. New York: Theatre Arts Books.

Sterne, L. (1980). *Tristram Shandy*. New York: Norton.

Toffler, A. (1990). *Power shift*. New York: Bantam.

Tosi, V. (1992). Marey and Muybridge: How modern biolocomotion analysis started. In A. Cappozzo, M. Marchetti, & V. Tosi (Eds.), *Biolocomotion: A century of research using moving pictures*. Rome: Promograph.

Trepman, E., Gellman, R.E., Micheli, L.J., & De Luca, C.J. (1998). Electromyographic analysis of grand-plié in ballet and modern dancers. *Medicine & Science in Sports & Exercise, 30(12)*, 1708-1720.

Trew, M., & Everett, T. (2001). *Human movement: An introductory text* (4th ed.). Edinburgh: Churchill Livingstone.

Unnithan, V.B., Dowling, J.J., Frost, G., & Bar-Or, O. (1999). Role of mechanical power estimates in the O_2 cost of walking in children with cerebral palsy. *Medicine & Science in Sports & Exercise, 31(12)*, 1703-1708.

Waters, T.R., Putz-Anderson, V., Garg, A., & Fine, L.J. (1993). Revised NIOSH equation for the design and evaluation of manual lifting tasks. *Ergonomics, 36(7)*, 749-776.

Whiting, W.C., & Zernicke, R.F. (1998). *Biomechanics of musculoskeletal injury*. Champaign, IL: Human Kinetics.

Williams, K., Haywood, K., & VanSant, A. (1991). Throwing patterns of older adults: A follow-up investigation. *International Journal of Aging and Human Development, 33(4)*, 279-294.

Williams, K., Haywood, K., & VanSant, A. (1998). Changes in throwing by older adults: A longitudinal investigation. *Research Quarterly for Exercise and Sport, 69(1)*, 1-10.

index

about the authors

William C. Whiting, PhD, is professor and director of the Biomechanics Laboratory in the department of kinesiology at California State University at Northridge, where he has won both the Distinguished Teaching Award and Scholarly Publication Award. Dr. Whiting earned his PhD in kinesiology at UCLA. He has taught courses in biomechanics and human anatomy for more than 15 years and has published more than 35 articles and 25 research abstracts. He is coauthor of *Biomechanics of Musculoskeletal Injury.*

Dr. Whiting currently serves on the editorial board of ACSM's *Health and Fitness Journal* and serves as a reviewer for a number of scholarly journals. Dr. Whiting is a fellow of the American College of Sports Medicine and has served as president of the Southwest Regional Chapter of ACSM. He is also a member of the American Society of Biomechanics; the International Society of Biomechanics; the National Strength and Conditioning Association; and the American Alliance for Health, Physical Education, Recreation and Dance

In his leisure time, Dr. Whiting enjoys playing basketball and volleyball, reading, camping, and hiking. He lives in Glendale, California, with his wife, Marji, and son, Trevor

Stuart Rugg, PhD, is an associate professor and chair of the department of kinesiology at Occidental College in Los Angeles. He received his doctoral degree in kinesiology, with an emphasis in biomechanics, from UCLA. For the past 17 years he has taught classes in human anatomy and biomechanics at Occidental College. Dr. Rugg has received Occidental's Outstanding Professor honor and is a four-time recipient of the college's Outstanding Teaching Award. His research focuses on the mechanical factors governing human performance and the effectiveness of sport equipment.

Dr. Rugg has taught a class in musculoskeletal anatomy and biomechanics for UCLA Extension's certified fitness training program and for the Mount Saint Mary's department of physical therapy. He is a member of the National Strength and Conditioning Association and has worked as a design consultant for exercise and sport equipment companies. Dr. Rugg is an accomplished nature photographer and enjoys reading, camping, hiking, rafting, cycling, and weightlifting.

essentials of interactive functional anatomy

Minimum System Requirements

PC

- Windows® 98/2000/ME/XP
- Pentium® processor or higher
- At least 32 MB RAM
- Monitor set to 800 x 600 or greater
- High-color display

Mac

- Power Mac®
- System 8.6 /9/OSX
- At least 64 MB RAM
- Monitor set to 800 x 600 or greater
- Monitor set to thousands of colors

How to Use This Program

PC

The program should launch automatically when the CD is inserted in the CD-ROM drive of your computer. Choose the Install button for IFA Essentials. If you don't already have QuickTime 6 installed on your computer, you should also install that. If the CD does not auto-launch, go to My Computer and double-click the HK_IFA_Ess icon. Install as stated above.

Mac

Insert the CD into the CD-ROM drive of your computer, then double-click the CD-ROM icon. Double-click the IFA Essentials for Power Mac folder, then the IFA Essentials icon. This will launch the program. If you don't already have QuickTime 6 installed on your computer, you should also install that using the QT installer in the IFA Essentials for Power Mac folder.

Quick Start Instructions

Use the Anatomy tab to view the 3D model and click on any anatomical structure to display the relevant text. Click on the red text for hot links to additional and relevant information about the chosen anatomical component. The History function provides access to previous text articles. Use the blue arrow buttons located in the upper right-hand corner of the text interface to move sequentially through the text. To maximize the model interface, place the cursor over the model/text interface until the double-arrow sign appears, left-click, and drag to the right. To maximize the text interface, drag to the left (PC version only). Rotate the 3D model by using the blue arrow buttons centered under it. The inner buttons rotate the 3D model step by step and the outer buttons rotate it continuously. Strip away anatomical layers, from deep to superficial, using the layer slider centered under the rotation arrows. Change the view of the 3D model by choosing additional views from the drop-down menu located on the lower right-hand side of the model interface.

To help you get started, Help balloons are available. Move the mouse over any button and its function will be revealed. (For a Mac, enable this by selecting Help balloons from the Help menu.) In-depth help is available from the Help menu.